OIL AND GAS, TECHNOLOGY AND HUMANS

Oil and Gas, Technology and Humans
Assessing the Human Factors of Technological Change

Edited by

EIRIK ALBRECHTSEN
SINTEF, Norway

&

DENIS BESNARD
Mines-ParisTech, France

CRC Press
Taylor & Francis Group
Boca Raton London New York

CRC Press is an imprint of the
Taylor & Francis Group, an **informa** business

CRC Press
Taylor & Francis Group
6000 Broken Sound Parkway NW, Suite 300
Boca Raton, FL 33487-2742

First issued in paperback 2017

© 2013 by Eirik Albrechtsen and Denis Besnard
CRC Press is an imprint of Taylor & Francis Group, an Informa business

No claim to original U.S. Government works

Version Date: 20160226

ISBN 13: 978-1-4094-5600-1 (hbk)
ISBN 13: 978-1-138-07473-6 (pbk)

Visit the Taylor & Francis Web site at
http://www.taylorandfrancis.com

and the CRC Press Web site at
http://www.crcpress.com

Contents

List of Figures

List of Tables

About the Contributors

Eirik Albrechtsen (ed.) is Senior Research Scientist at SINTEF Safety Research, Norway and an Associate Professor in Safety Management at the Norwegian University of Science and Technology (NTNU). He has worked on various research projects on safety and Integrated Operations (IO) and has written several scientific publications on the risk of major accident and IO. He holds a PhD in Safety Management from NTNU (2008).

Denis Besnard (ed.) is a Research Associate at MINES ParisTech, France. He has written several articles, book chapters and reports on the human contribution to systems safety. He is the scientific director of a French executive post-Master's degree on the Human Factors of Safety Management. He also carries out safety management consultancy in the industry. He holds a PhD in Psychology from the University of Provence (1999).

Siri Andersen is currently a PhD student at the NTNU studying operational risk assessment in an IO context. She has several years of experience as a risk consultant at Det Norske Veritas, where she mainly worked in the petroleum industry.

Kari Apneseth is a Safety Technology Engineer in the Health, Safety and Environment (HSE) Competence Centre at Statoil. She has previously worked in HSE performance and monitoring. She holds a Master's degree in HSE from the Department of Industrial Economics and Technology Management at the NTNU, with a specialization in Safety Management.

Alf Ove Braseth is a principal scientist at the Institute for Energy Technology in Norway. He is a product designer mainly working on information visualization on large screen displays and holds several design patents. He has an MSc in Mechanical Engineering.

Tor Olav Grøtan is a Senior Research Scientist at SINTEF Safety Research, Norway. He holds a Master's degree in Engineering Cybernetics and another in Management. He has been working on the organizational aspects of safety in the military, healthcare, transportation and petroleum industry domains. IIe is currently a PhD student at the NTNU.

Stein Haugen is a Professor in Technical Safety at the NTNU, where he mainly works in the areas of risk analysis and prevention of major accidents. He carried out consultancy work and research projects in the industry for more than 25 years before joining the University.

Erik Hollnagel is Professor at the University of Southern Denmark, Chief Consultant at the Centre for Quality in the Southern Denmark Region, Industrial Safety Chair at MINES ParisTech, France and an Associate Professor at the NTNU. He is recognized internationally as an expert in resilience engineering and safety management of complex socio-technical systems.

Jan Hovden is a Professor in Safety Management at the NTNU. He has written several articles and reports on societal safety as well as safety management in various industrial sectors. He is a member of several national and international expert advisory groups.

Grete Rindahl is a principal scientist at the Institute for Energy Technology in Norway. She has been leading research projects in IO for several years, focusing on issues related to visualization and collaboration technology, IO training and IO teamwork. She holds a Master's degree in Applied Mathematics from the University of Tromsø.

Sizarta Sarshar is a Senior Research Scientist at the Institute for Energy Technology in Norway. He holds an MSc in Computer Science with a focus on safety-critical systems and is currently working on his PhD. The topic of his work is decision making in integrated operation collaboration processes, with a focus on visualizing safety hazard indicators in offshore operations planning for the prevention of major accidents.

Ann Britt Skjerve is a Principal Scientist at the Institute for Energy Technology in Norway. She has worked on several research projects related to training, teamwork and system usability in the domains of nuclear power plant operation and IO in the petroleum industry. She holds a PhD in Psychology from the University of Copenhagen.

Camilla Knudsen Tveiten is currently a consultant at Proactima AS within risk management and HSE support. She is also partly working on her PhD at the NTNU on conditions for resilient operations of complex systems undergoing technological alternations. She has several years of experience in the petroleum industry and in research in SINTEF within the HSE and risk management area.

Jørn Vatn is a Professor at the NTNU. His research interests include risk analysis, reliability analysis and maintenance optimization. He has published several scientific papers on risk assessment and the interpretation of risk and has participated in several groups dealing with both risk assessment and risk communication.

Aud Marit Wahl is currently a Senior Engineer with the Norwegian National Rail Administration. She has been employed as a researcher with MARINTEK (part of the SINTEF group), working on projects on IO and planning in the oil and gas industry. She holds a Master's degree in Social Sciences from the NTNU.

Acknowledgements

This book is the product of the 'Interdisciplinary Risk Assessment in Integrated Operations addressing Human and Organisational Factors' (RIO) research project, which ran from 2008 to 2012. It was sponsored by the Norwegian Research Council's PETROMAKS programme, the Norwegian Petroleum Safety Authority and the Center for Integrated Operations in the Petroleum Industry at the Norwegian University of Science and Technology (NTNU).

The editors would like to thank Elaine Seery (http://science-proofreading.net/) for language editing and preparation of the manuscript.

Chapter 1
Introduction and Overview

Eirik Albrechtsen and Denis Besnard

The term Integrated Operations (IO) was introduced at the start of the twenty-first century. It was coined when part of the bandwidth of a telecom cable was used by an oil company in the North Sea for closer onshore/offshore operations. Today, IO refers to new working practices and technologies made possible through the introduction of high-bandwidth communication technology (particularly fibre optics) in the offshore oil and gas industry. IO can be defined as the integration of people, work processes and technology intended to improve decision-making and operational performance. IO feature ubiquitous real-time data, collaborative technologies and cross-discipline expertise provided by different organizations in various geographical locations.[1] The aim of IO is to improve the safety of both operations and decision-making and it would therefore seem to offer a promising way to organize work. However, oil and gas production is a risky industry where major accidents can cause severe human, material and environmental losses. Consequently, the adoption of any new technology or working practices means that change must be managed, as must the risks related to change. Managers need to be made aware of these opportunities and challenges; that was the starting point for this book.

Although it would be reasonable to reassess traditional risk assessment methods in the context of the changes brought about by the introduction of IO, recent fieldwork paints a different picture. In an article that predated this book by just a few years, two of

---

1   This definition is used by the Center for Integrated Operations in the Petroleum Industry, Trondheim, Norway.

our fellow authors (Andersen and Mostue, 2012) argued that the existence and extent of changes provoked by the introduction of IO is not always acknowledged in daily operations. Their interview-based study pointed out that new working practices in the industry have influenced risk assessment in three ways:

- First, there is a belief that existing risk assessment methods can be applied to new working practices and technologies. However, in reality the data that forms the basis for risk assessment has changed, which means that experts from a more diverse range of disciplines must be involved.
- Second, field interviews indicated that current methods were inadequate for the assessment of human and organizational factors, which also supports the idea of involving personnel from other disciplines.
- Third, day-to-day risk assessment operations tended to deploy formal and informal resilience-based approaches (for example, anticipating and monitoring current events) rather than specialized tools or methods.

The adoption of IO may have a more serious impact on risk management. One example – optimism – was identified by Knudsen (2010). In a survey of industry personnel, respondents thought that IO features such as real-time data, visualization technology and closer cooperation between actors would improve risk management. However, these new working practices would in fact require new risk assessment and management methods (Albrechtsen et al., 2010). These methods include the assessment of intractable systems, complex collaboration patterns and human–machine interactions (HMIs).

The introduction of IO raises several important questions. How should the consequences of change be managed? What is the best way to assess risks associated with the deployment of a technology that may have an impact on entire divisions of the organization? In such a context, what is the appropriate level of understanding needed to assess risks? Which risk assessment techniques are most suitable for the successful management of change? This book attempts to answer these questions and provide an overview of risk assessment concepts and techniques that are relevant to IO safety managers, regulators, risk researchers and other stakeholders in the oil and gas industry.

Where It All Started: The RIO Project

The contents of this book were produced within the 'Interdisciplinary Risk Assessment in Integrated Operations addressing Human and Organisational Factors' (RIO) research project. The RIO project aimed to develop a better understanding of how to assess risk associated with the changes brought about by the introduction of IO in the Norwegian oil and gas industry. A requirement of the project was to establish guidelines based on theory, modelling and logical frameworks.

The project took an interdisciplinary approach and drew upon expertise from the disciplines of risk assessment, safety management, psychology and Resilience Engineering. Consequently the contents of this book are also the result of interdisciplinary collaboration and provide a variety of perspectives related to decision support for the prevention of major accidents. The various chapters describe, and in some cases combine, contributions from different disciplines. We reflect on this point in the concluding chapter, which discusses the strengths and weaknesses of an interdisciplinary approach and the lessons learned.

The Need for an Interdisciplinary Approach

There are various reasons why it makes sense to use an interdisciplinary approach to understanding and assessing IO-related risk. First, IO itself promotes interdisciplinary working practices. For example, drilling expertise centres are composed of teams of experts from different disciplines, such as drilling and subsea operations, drilling technology, geology, well control and completions, and rock mechanics. Second, in socio-technical systems safety issues cross-cut activities. Knowledge from various activities is combined in order to fully understand the overall system. Third, managing the risk of a major accident requires more than one skillset. For instance, Rasmussen (1997) argues that low-frequency, high-severity events should be addressed using an interdisciplinary risk-based approach. Also, March et al. (1991) highlight the benefits of encouraging interdisciplinary collaboration and argue that an interdisciplinary approach is needed in order to better learn from and interpret rare events. Although this could be said to be true of many projects, it is particularly useful in the management of safety-critical socio-technical systems, whether in the Norwegian petroleum industry or other safety-critical industries. An interdisciplinary approach offers significant benefits in capturing the consequences of change. When applied to risk assessment, it can alert managers to areas where risks may arise and suggest ways to approach their evaluation.

This book avoids a techno-centric approach to risk assessment. We did not want to provide an engineering manual focused simply on the deployment of risk assessment tools, without any explanation of choices, assumptions and alternatives. It seems to us that this approach introduces a bias, in that it tends to exclude the surrounding organization. Although this book describes methods, the focus is on humans and their working environment. After all, risks do not arise from the introduction of technology alone. They are rather the result of a combination of the way humans use technology, the decisions that were taken about its adoption and use, and the operational culture created by an organization. Readers are left to judge the added value that this approach offers for risk management.

More practically, bringing scientists from different backgrounds together in the same room to collaborate is a challenge in itself. This was equally true in the preparation of this book and it is probably true of any industrial project where decisions must be taken. A few reasons for this state of affairs can be identified. For example, although putting aside a narrow focus on one's own discipline and listening to others clearly facilitates collaboration, this is not how academia is organized. Researchers are encouraged to become specialists in their discipline and, over time, specialization turns knowledge into a very selective lens through which the world is seen and understood. This is one of the drawbacks of expertise. However, an interdisciplinary team sees the world as an object that cannot be understood other than through a variety of lenses. The challenge is then to actually get people to work together. One effective way to achieve this is to focus on a tangible problem. In this book, that problem was IO.

The Audience for This Book

This book could have been a manual. Manuals are useful because they tell people what to do. However, they do not always explain why certain actions must be taken, nor do they necessarily highlight any limitations or assumptions. Therefore, this book explains the basics, outlines options and highlights techniques. This makes it less immediately useful than a manual as it does not contain ready-made solutions to problems. On the other hand, it provides the knowledge required to understand a problem rather than offering a solution to an issue that might not be fully understood. Consequently, the reader is able to take informed decisions founded on a better understanding of the overall context.

This may seem a strange decision given that a) our intended audience is risk managers and b) risk managers typically have a chronic lack of time. Therefore, some explanations are in order. First, we think that an investment in understanding the foundations of a given class of problems yields a better return than the blind deployment of a method. This is an important point. We know from experience that safety management methods are sometimes chosen on the basis of their popularity rather than their suitability to deal with the problem at hand. Second, the book does document methods and their deployment in a reasonable amount of detail although the focus is always on assumptions and limitations. Finally, an important motivation for this book was the transversal nature of IO and consequently the potentially diverse readership. There are many similarities between the introduction of IO in the oil and gas industry and in other, very different fields such as telemedicine, remote control of transportation and energy supply systems, and automation of the control and supervision of production chains. We are therefore hopeful that the concepts, challenges and methods described in this book can be applied in other domains.

Overview

The book takes a progressive approach to IO. It first describes the main concepts and definitions, then documents and discusses the deployment of risk assessment methods. Although many chapters are self-contained and can be read individually, starting at the beginning will give readers a broader understanding of IO and risk assessment, and progressively build a multifaceted understanding of the issues. The book is structured into three main sections:

- Foundations;
- Operations and Risk Assessment; and
- Risk Assessment of an IO Scenario from Different Perspectives.

The Foundations section consists of Chapters 2–5. In Chapter 2, Eirik Albrechtsen gives an overview and a definition of IO that shows the extent to which the introduction of IO can impact an organization. Albrechtsen shows how new technology makes it possible to combine data from various locations and domains of expertise. However, technology alone cannot compensate for the inherent limitations of an organization, such as poor working arrangements, insufficient communication, lack of skills and so on. This point is developed by Siri Andersen in Chapter 3, who provides an overview of potential hazards in an IO context. Furthermore, this chapter describes a diagnostic checklist to help in the audit of new working practices in drilling activities. The third contribution to this section comes from Tor Olav Grøtan. This chapter discusses a potential outcome of IO, namely complexity. IO changes the way work is done and crosses geographical and discipline boundaries. Grøtan argues that three complexity-related issues must be taken into account when assessing risk: the organization of attention, sensemaking, and the remaking of the organization. Finally, Chapter 5 summarizes the lessons that can be drawn from this first section.

Chapters 6–11 form the section on Operations and Risk Assessment. Chapter 6 by Jørn Vatn and Stein Haugen discusses risk processing and its role in the acquisition of safety-related knowledge. Using examples from recent industrial catastrophes, they question whether certain risk assessment methods are appropriate for the prediction of rare events. Chapter 7 is also method-oriented. In this chapter Denis Besnard focuses on HMIs in remotely-operated drilling operations and describes the deployment of a risk assessment method, from the theoretical building blocks to its production and use. The philosophy is echoed in Chapter 8 by Kari Apneseth, Aud Marit Wahl and Erik Hollnagel, who describe the assessment of Integrated Planning (IPL). This feasibility study documents the methodological stages involved in the adaptation and deployment of a resilience assessment method. Camilla Tveiten adopts a similar approach in Chapter 9, which describes the implementation of a functional resonance risk assessment method (FRAM) that was used to assess the impact of variability in the planning of oil and gas production activities. With this method the risk assessment exercise could be carried out at organizational level. Finally, Chapter 10 by Grete Rindahl, Ann Britt Skjerve, Sizarta Sarshar and Alf Ove Braseth addresses hazard identification in maintenance planning using a computer-driven collaboration tool. Their chapter describes the results of a field survey and highlights how the tool facilitated hazard identification. Chapter 11 summarizes the lessons learned from this second section.

The third and final section deals with Risk Assessment of an IO Scenario from Different Perspectives. The objective of this section is to compare two risk assessment approaches. In Chapter 12 Eirik Albrechtsen outlines a sample IO scenario. The scenario was designed to explicitly highlight various system-wide, organizational dimensions of risk. In Chapter 13 Jørn Vatn presents one risk assessment approach. This chapter shows how qualitative factors such as communication between stakeholders, understanding of assumptions, and verification can be integrated into a quantitative assessment. In Chapter 14, Erik Hollnagel describes an alternative, resilience-based analysis method. In this chapter the impact of the introduction of IO is assessed according to four organizational abilities: learning, responding, monitoring and anticipating. Finally, in Chapter 15 Eirik Albrechtsen and

Denis Besnard compare the two approaches and ·reflect on the strengths and weaknesses of each. Chapter 16 summarizes the lessons learned from this third and final section.

Chapter 17 offers some conclusions. In this chapter, Denis Besnard, Eirik Albrechtsen and Jan Hovden step back from technical issues and highlight some of the practical implications of the introduction of IO for risk assessment.

References

Albrechtsen, E., Andersen, S., Besnard, D., Grøtan, T.O., Hollnagel, E., Hovden, J., Mostue, B.A., Størseth, F. and Vatn, J. 2010. *Essays on Sociotechnical Vulnerabilities and Strategies of Control in Integrated Operations*. Technical report SINTEF A14732, SINTEF-NTNU, Norway.

Andersen, S. and Mostue, B.A. 2012. Risk analysis and risk management approaches applied to the petroleum industry and their applicability to IO concepts. *Safety Science*, 50, 2010–2019.

Knudsen, R.H. 2010. *Myter og sannheter om risikoanalytisk tilnærming* [Risk assessment – truths and myths]. Master's thesis. Trondheim, Norway: Norwegian University of Science and Technology.

March, J.G., Sproull, L.S. and Tamuz, M. 1991. Learning from samples of one or fewer. *Organization Science*, 2, 1–13.

Rasmussen, J. 1997. Risk management in a dynamic society. A modelling problem. *Safety Science*, 27, 183–213.

Section One
Foundations

Chapter 2
Integrated Operations Concepts and Their Impact on Major Accident Prevention

Eirik Albrechtsen

New ways of doing and organizing work based on the application of Information and Communication Technology (ICT) are being used and developed in the offshore oil and gas industry. The aim is both to increase value creation and reduce the risks of a major accident. These new ways of doing work have been named Integrated Operations (IO) by some of the main actors in the industry. IO can be understood as the integration of people, work processes and technology in order to take smarter decisions and ensure better execution. This is enabled by the use of ubiquitous real-time data, collaborative techniques and expertise that crosses disciplines, organizations and geographical locations.

These various aspects of IO introduce both new possibilities and challenges in dealing with major accident risks. Some benefits of IO are that it supports risk-informed decision-making (for example, by improving the quality and relevance of safety-related data, including real-time data), provides more information, offers better ways to visualize risk information (more information is available and can be accessed anywhere – anytime) and provides access to experts in multidisciplinary onshore support teams. On the other hand, inadequate IO solutions may be a factor that contributes to major accidents, through for example poor collaboration and integration of onshore–offshore teams and lack of information flow between relevant actors.

Introduction

On the Norwegian Continental Shelf, there is a sustained trend towards the development, deployment and application of ICT and digital infrastructure that creates new ways of organizing and executing work. The petroleum industry has declared that this application of ICT is both a prerequisite and a driving force for the development of better integrated work processes in all main areas of activity, such as drilling operations and maintenance. The initiative began at the start of the current millennium and is closely linked to the aim of increased value creation through (OLF, 2003):

- efficient reservoir exploitation;
- optimization of exploration and operation processes;
- ambitions for the long-term development of fields and installations; and
- improved Health, Safety and Environment (HSE) performance.

These developments have been named Integrated Operations (IO) by the Norwegian Oil and Gas Association and some of the main operators on the Norwegian Continental Shelf. The Norwegian Center for Integrated Operations in the Petroleum Industry has defined IO as, 'The integration of people, work processes and technology to take smarter decisions and better execution. It is enabled by the use of ubiquitous real-time data, collaborative techniques and multiple expertise across disciplines, organizations and geographical locations.' Other companies use similar concepts such as intelligent fields, smart fields, field of the future and iFields. However the basic idea remains the same. The idea has become so commonplace that some actors in the industry now claim that they do not talk explicitly about the notion of IO anymore; it has been incorporated into the 'way they work'.

This chapter identifies some generic properties of IO in the offshore oil and gas industry. This is followed by a description of some of the new ways of doing work in an IO context and how they may affect safety. Finally, the chapter explores some of the implications for major accident prevention in an IO context, showing both positive and negative effects.

Integrated Operations Concepts

The purpose of introducing IO is value creation through smarter, better and faster decision-making and execution. Lilleng and Sagatun (2010) argue that seven interdependent conditions need to be in place to enable this value creation:

- *New technology* enabling new methods of data capture that were previously too costly and risky.
- *Increased capacity in the communication infrastructure*, for example, via fibre optics, to handle accelerated data capture across geographically distributed actors.
- *Integration and processing of data* from various sources that makes information easy to access by users across disciplines and companies.
- *Presentation and visualization of information* for everyone who needs it in a user-friendly manner.
- *Interdisciplinary collaboration work arenas* where team members can connect to other members and have access to information, for example, collaboration rooms for various offshore as well as onshore units.
- New ways of *organizing work* processes based on information distribution and collaboration arenas that a) integrate operators, contractors and service companies more closely; b) integrate onshore and offshore organizations; and c) establish expert support. Work is distributed across geographical, organizational and disciplinary borders.
- An *IO mindset* among staff that enhances the possibilities provided by technology, for example, trust and openness to knowledge sharing, interdisciplinary collaboration and focus on continuous change management.

Central to the idea of increased value creation is the expectation of more effective decision-making (OLF, 2003, Ringstad and Andersen, 2007). The main driving forces behind IO are the optimization of exploration and operations, reduced costs and improved HSE performance. Henderson et al. (2012) presented a capability platform (a set of interdependent capacities and abilities involving people, processes, technology and governance) to show how IO can create value. They differentiate between three types of IO capabilities:

- *Foundational capabilities.* Core technology resources (for example, the equipment needed to move hydrocarbons from the reservoir to the market) must still be in place in an IO setting. To enable IO solutions, safe and reliable data communication and infrastructure must exist to both collect and distribute data.

- *Analytics and collaboration capabilities.* To make use of the collected data, real-time processing is required, which makes the data available as support for decision-making. Sharing information and knowledge across disciplines is necessary.
- *Operational capabilities* are the development of new work processes and support for decisions that realize the business potential of IO, including major accident prevention.

For each IO capability to produce value, the sub-capabilities related to technology, people, processes and organization must be in place (Henderson et al. 2012).

Both Lilleng and Sagatun (2010) and Henderson et al. (2012) compile IO conditions/capabilities in stack models. The idea of a stack is that all levels in the model need to be in place to enable value creation. For example, data collection and distribution by technological resources (the lower levels of the stack) need to be in place to enable higher-level properties such as visualization and distributed work arenas. The two stack models are shown in Figure 2.1 demonstrate how the layers of the models are interlinked. The three bottom layers in the IO element stack are important factors that need to be in place in order to generate the foundational capabilities shown in the IO capability stack on the right-hand side of the figure. Likewise, the elements in light grey in the element stack model contribute to enable analytical and collaboration capabilities. The two topmost elements in the left-hand model (among others) support operational capabilities, which appear in dark grey in both stack models.

Taken literally, IOs are about integration. This means that the various layers and capabilities shown in Figure 2.1 must be integrated; similarly each layer in the model must be internally integrated. Furthermore, the same idea applies to each element of the value chain and employers, disciplines and organizations.

The next subsection illustrates some new ways of working in an IO context. It also highlights some opportunities and challenges for safety in the context of IO-based solutions.

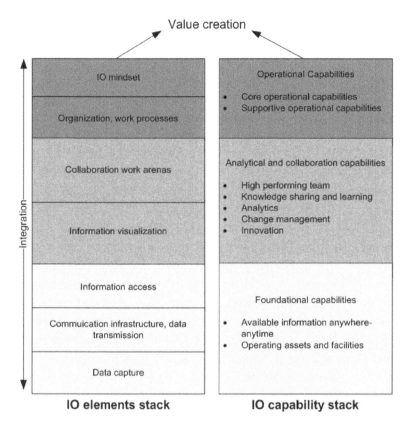

Figure 2.1   IO elements stack model (based on Lilleng and Sagatun, 2010) and the IO capability stack (based on Henderson et al., 2012)

Examples of Integrated Operations-based Solutions

Technology as an enabler for automating and digitizing work: integrated drilling simulators

New technology both enables and is a requirement for IO development. The immense capacity of fibre optic communication is a precondition, since most IO development is, in some way, based on increased bandwidth (OLF, 2005). As the stack models in Figure 2.1 show, digital infrastructure makes different IO solutions possible through technological capabilities such as:

- the capture, visualization and transfer of real-time data;
- data standardization and integration;

- integration of applications, including integration of ICT systems and process control systems;
- virtual collaborative arenas; and
- automation and instrumentation of work processes.

These technological IO solutions impact work processes, either by automating or by applying information to perform work (Zuboff, 1988). Information technology (IT) both automates manual activities and provides support by access to information, which can be used to understand, improve and plan activities.

Integrated drilling simulators are one example of an IO solution that applies real-time information to improve work practices. The eDrilling concept is an example of such a system. It belongs to a class of systems that enable remote control/ support by an onshore drilling expert centre using means such as real-time data monitoring, visualization and data simulation (Rommetveit et al., 2008). Real-time data makes an integrated drilling simulator possible as it mirrors the drilling process itself and provides information on key parameters to onshore support centres. These onshore centres, consisting of various experts, use the information provided to support the offshore drilling team. This enables safer and more efficient drilling operations. Such a system can also automatically diagnose upcoming problems using real-time simulations.

Figure 2.2 illustrates how access to data and data sharing connects different actors and enables onshore real-time monitoring of offshore operations. Improved and more up-to-date data is collected offshore by intelligent equipment. This data can be shared among offshore and onshore actors, and operators and vendors are able to process, analyse and visualize real-time data.

Integrated drilling simulators provide several opportunities for safer drilling operations (Rommetveit et al., 2008). These include: real-time supervision of the drilling process; diagnosis of the drilling state and conditions; early warning of upcoming unwanted conditions and events; forward simulations through tests of drilling plans as well as what-if evaluations and training possibilities.

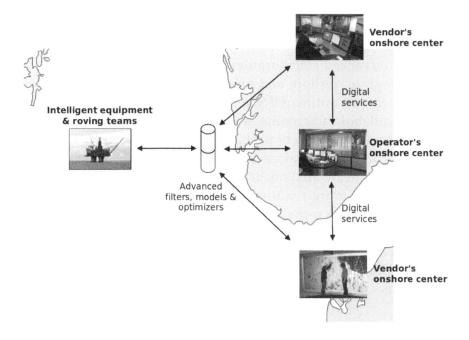

Figure 2.2   Collection, processing and distribution of real-time data enables new ways of organizing work

Source: Figure redrawn from Norwegian Oil and Gas Association (OLF, 2005). Pictures of onshore centers: The Centre for Integrated Operations

However, the application of the digital infrastructure can also introduce new vulnerabilities and threats. One of the most critical aspects of technological integration on the Norwegian Continental Shelf is the merger of offshore process control systems with web-based onshore administrative ICT systems. This infrastructure introduces the opportunity for breaches of information security that can be an obstacle to IO-based ways of operating (Johnsen, 2008, Johnsen et al., 2008). It also means that the digital infrastructure is one of the most critical information infrastructures in Norway. For these reasons, not only is information security a business-critical issue for the oil and gas sector, it also imposes security requirements and the prevention of major accidents being integrated in an IO context.

Ways and means for communication and collaboration in an IO context: integration of offshore and onshore operations

New and advanced collaboration technology makes it possible to better integrate onshore and offshore personnel. Most assets on the Norwegian Continental Shelf have implemented onshore and offshore collaboration rooms, see Figure 2.3. The aim is to develop a shared understanding of problem-solving between onshore and offshore teams and obtain more hands-on support from onshore experts. This means that onshore management teams and domain experts can be more readily and actively involved in problem-solving and decision-making related to offshore operations.

Offshore, many processes have and will remain mostly unchanged despite the IO solutions implemented by organizations. The installation's management team remains responsible for operations (with support from the onshore management team).

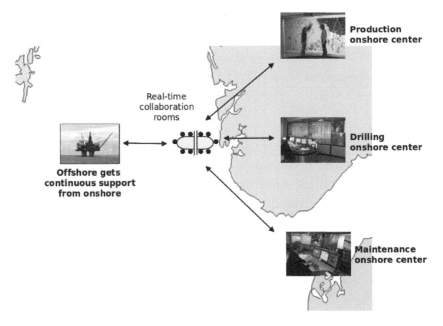

Figure 2.3   Collaboration technology makes it possible for onshore centres to provide support to offshore operations

Source: Figure redrawn from Norwegian Oil and Gas Association (OLF, 2005). Pictures of onshore centers: The Centre for Integrated Operations

However, some administrative and operational tasks can be transferred from offshore to onshore. Such changes address both safety concerns (for example, reduced manning, less need for helicopter transport) and the need for a lean organization. However, the transfer of functions from offshore to onshore has meant that a number of onshore activities have become more important in ensuring operational safety offshore. According to Skarholt et al. (2009) and Skarholt and Torvatn (2010) the cultural differences between onshore and offshore organizations have been reduced in IO environments. They also indicate that the use of collaboration rooms has led to improvements in the coordination and communication of tasks as well as quicker responses from onshore centres to requests from offshore.

The technology that supports this collaboration consists of high-quality video conferencing and information sharing interfaces. The belief is that these technologies reduce contextual differences (working environments, weather conditions and so on) between offshore and onshore teams (Skarholt et al., 2009). However, whether shared situational awareness is really established between onshore and offshore personnel is a disputed issue. In particular, concerns have been raised about the potential future increase in the lack of offshore experience in onshore teams.

The application of collaboration technology and shared information not only integrates offshore and onshore personnel, it also enables integration across organizational borders. An example is the ability to integrate operating companies and contractors, for example in morning meetings where contractors participate using collaborative technology.

Access to onshore, interdisciplinary expert knowledge: competence integration

The previous subsection showed how offshore operations and onshore support can be more closely interconnected. A central factor in improved onshore decision-making support is interdisciplinary expert centres, for example, logistic planning centres. Collaborative electronic interfaces provide access to broad repertoires of knowledge, resources and expertise that can support both routine and crisis situations. Herbert et al.

(2008) describe how not only competencies and disciplines, but also service providers are integrated in such an onshore drilling support centre. Many disciplines can be centralized in these centres, including operational geologists, data or logging engineers, drilling optimization engineers, directional drillers and others. The centre can support four to six drilling operations simultaneously. For instance, Herbert et al. (2008) studied such a support centre and identified the safety profit of its implementation. During its first five years of operation, they estimated that over 900 helicopter seats were saved and that 9,000 fewer days were spent offshore.

Skills integration offers the personnel and experts involved in the various phases of oil and gas production a shared frame of reference. Competence integration brings together the various skills and knowledge held by stakeholders involved in the various processes (for example, drilling and well operations, and reservoir, well, process, condition and HSE understanding). This interdisciplinary decision-making support is a key property of IO.

Weltzien (2011) and Albrechtsen and Weltzien (2012) show how such an interdisciplinary expert centre for offshore drilling can facilitate safer drilling operations. Sharp-end drilling activities require constant adaptation to dynamic operations. These studies show that the onshore support centre provides important support for these sharp-end adaptations due to the centre's abilities to monitor and foresee future trends. The onshore centre acts as a second pair of eyes for drillers by monitoring what is going on in the drilling operation. Real-time data combined with multidisciplinary teams improve the ability to detect anomalies and trends. Additionally, the onshore team can perform in-depth analysis of problems and provide recommendations for the offshore team. The onshore support team are able to identify the warning signs of imminent phenomena and situations and can allocate resources to increase monitoring and prepare responses. Teams can then adapt to situations by focusing on problems they suspect may happen and prepare solutions to cope with the related challenges.

The various premises and underlying qualities of an IO context provide the conditions for the abilities described in the previous paragraph (Weltzien, 2011). Working in interdisciplinary teams in open-plan offices creates an environment that encourages collaboration, shared ideas, and knowledge and redundancy. Shared information interfaces provide further support as teams can view and access the same information. The application of statistical and mathematical analysis to real-time and historical data can provide valuable decision support during planning and operations. Reports from operations, drilling projects and risk assessments are documented and can be shared within the whole organization.

Major Accident Prevention within Integrated Operations

The previous subsection showed that the introduction and application of IO solutions creates both opportunities and challenges for the prevention of major accidents. Reviews of recent major events in the industry indicate some potential challenges related to the introduction of IO. This section begins with an assessment of the IO aspects of recent blowouts in the offshore oil and gas industry. Next, improvements to risk assessment using IO-based solutions are discussed.

Integrated Operations-related Aspects of Recent Blowouts

The last decade has seen some major accidents and near-accidents in the offshore oil and gas industry. These include the Deepwater Horizon accident in the Gulf of Mexico in 2010, the Montara blowout off the coast of Australia in 2009, the gas blowout and near-accident at Snorre A in 2004 and the Gullfaks C well incident in 2010. The latter two events both occurred on the Norwegian Continental Shelf. Tinmannsvik et al. (2011) reviewed the investigation reports prepared following the above-mentioned incidents, with particular emphasis on the Deepwater Horizon accident. This review indicates that although IO is not mentioned explicitly in any of the investigation reports, there are factors that contributed to events that can be related to IO solutions.

Sharing Information

One of the aims of implementing IO is to create environments for the efficient sharing of information between several actors and improved presentation of, and simpler access to information. An inadequate flow of information was a major contributing factor to the blowouts mentioned above. The Deepwater Horizon investigation report shows that poor communication between the actors involved together with a lack of shared information were among the major factors contributing to the accident (for example, Chief Counsel's Report, 2011). For example, the BP onshore team was aware of the increased risk caused by cementing, but did not inform offshore crew members about this increased risk (Chief Counsel's Report, 2011). In the Gullfaks C incident, the transfer of experience on measured pressure drilling was inadequate. Moreover, earlier events had occurred during the planning of the well operation (PSA, 2010, Statoil, 2010). In the Montara event, the Montara Commission of Inquiry (2010) identified poor information flow between night and dayshifts and onshore and offshore teams operating at the well in the period before the blowout.

Expert Support

Collaboration technology can enable onshore experts to provide critical support for offshore decisions. Onshore interdisciplinary expert teams can support sharp-end activities. However, the report of the Gullfaks C incident investigation showed that measured pressure drilling experts were not sufficiently involved in the planning, risk assessment and operational follow-up of the well operation (PSA, 2010, Statoil, 2010). Similarly, in the period before the gas blowout at Snorre A, experts were insufficiently involved in risk assessment and training (Schiefloe and Vikland, 2005). A similar situation was seen in the Deepwater Horizon accident: no expert was asked to assess abnormal data from a negative pressure test (Chief Counsel's Report, 2011).

Onshore/Offshore Collaboration

The two contributing factors described above point to a third issue: integration of onshore and offshore teams, which is a significant aspect of IO. In the Deepwater Horizon accident, the Chief Counsel's Report (2011) showed that BP had inadequate procedures to describe when offshore personnel should contact onshore support teams. Schiefloe and Vikland (2005) interviewed personnel involved in the Snorre A gas blowout and found poor levels of collaboration between offshore and onshore teams due, in part, to a lack of understanding of offshore conditions by onshore workers.

IO promises the improved capture, processing and visualization of real-time data and the provision of information anywhere and at any time. These are the bottom levels of the stack model shown in Figure 2.1. The stack model also indicates that for value creation this data and information must be interpreted and then used to support decisions. A key question following the Deepwater Horizon accident is why the drilling crew and the mud logger did not react to anomalous data and kick signals for nearly 50 minutes before the explosion. The Chief Counsel's Report (2011) indicated that lack of crew awareness, simultaneous operations and poor human–machine interfaces contributed to the failure to detect the signals. This demonstrates that greater access to data and information does not necessarily result in improved decision-making.

These examples show that inadequacies in IO-based solutions (among other factors) may have contributed to recent blowouts in the offshore oil and gas industry. However, on the other side of the coin, adequate IO solutions could have helped to prevent these incidents.

Integrated Operations Solutions to Improve Safety and Risk Assessments

More positively, various IO concepts can be used to support risk-informed major accident prevention. The two stack models shown in Figure 2.1 highlight the main elements of IO. These models can be used to show how various IO solutions can improve aspects of risk management. Figure 2.4 shows the connections between IO stack models and steps in risk management.

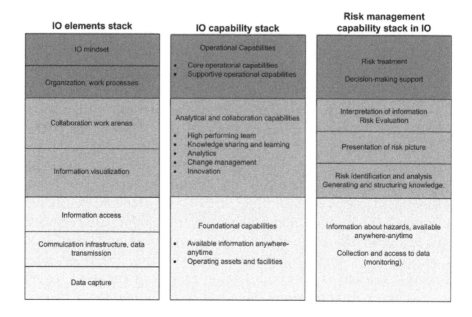

Figure 2.4   IO elements stack model (based on Lilleng and
              Sagatun, 2010), IO capability stack (based on
              Henderson, Hepsø and Mydland, 2012) and a risk
              management stack model

This book focuses on methods and approaches to assessing risk
in an IO context, which is mainly related to the risk identification
and analysis layer of the stack model shown in Figure 2.4. In this
context, Chapter 10 (Rindahl et al.) shows how risk information
can be visualized. Other risk management gains offered by
IO include an improved knowledge base; shared information
surfaces giving more people access to the same safety-related
information; and multidisciplinary teams that can collectively
interpret information and thereby provide improved decision
support.

There is another way in which IO can influence risk and safety
assessments. The new ways of doing and organizing work created
by the IO context requires the identification of new inputs to risk
assessment. These inputs are parts or combinations of the contents
of the two IO stack models shown on the left of Figure 2.4. In this

book we sketch out two main approaches to the assessment of new ways of doing and organizing work in an IO context:

- *Apply existing assessment methods but look for new input data.* Andersen (Chapter 3) identifies various hazards related to the organizational and human factors generated by IO-based solutions. This chapter also provides a checklist for the identification of these hazards. Vatn and Haugen (Chapter 6) discuss how different types of risk assessment methods can be improved, particularly those related to operative analysis. Vatn (Chapter 13) shows how the basic steps of a strategic quantitative assessment can address new IO-related factors.
- *Develop/apply new methods suited to an IO context.* Grøtan (Chapter 4) shows how different types of complexities brought about by the introduction of IO can be identified and managed. Both Apneseth et al. (Chapter 8) and Tveiten (Chapter 9) show how new assessment methods can be applied in an IO context. Finally, Hollnagel (Chapter 14) demonstrates how Resilience Engineering can be used as a basis to assess risk in IO.

Similarities to Other Sectors

Larsen et al. (2012:2) claim that the definition of IO (similar to the one used in this chapter) 'is by no means limited to the petroleum industry, and works very well when addressing integrated operations in the other sectors'. Their study shows the similarities between IO in the oil and gas industry and the military sector, medicine and aviation. The authors identified an issue common to these four sectors, namely the collection, processing and presentation of information to a group of people responsible for monitoring, control and safe execution. They showed that telemedicine is similar to IO in that it focuses on a distributed decision-support system involving experts and non-experts. Network-centric warfare that involves running complex and time-critical operations in the military sector shares many similarities with contents of the stack models shown in Figure 2.1. Such military operations use ICT to improve collaboration between units and create a common situational awareness. Communication between actors and experts distributed across various centres is a key advantage in such operations. There are various other sectors where ICT is used to enable elements and capabilities similar to those shown in Figure 2.1, for example, the aerospace, nuclear and power supply industries.

The similarities between the oil and gas industry and other sectors in the use of IO indicate that these sectors may provide the potential for risk management knowledge transfer. It also suggests that the ideas described in this book may be applicable to these sectors.

Conclusion

IO can be understood as new work practices in the offshore oil and gas industry that are made possible by use of ICT and digital infrastructure.

The generic properties of IO are:

- application of digital communication infrastructure;
- the capture of offshore performance data;
- the use of real-time data to monitor and manage operations across geographical and organizational borders;
- information that is available anywhere, anytime;
- the user-friendly presentation and visualization of information to everyone who needs it;
- interdisciplinary collaboration arenas;
- access to expert knowledge;
- integration of data, skills, activities and organizations.

The general belief is that these IO concepts will provide better, faster and safer decisions. In turn this will result in reduced operational costs, a longer life-span, accelerated and increased production, improved HSE performance and a reduced risk of major accident. However, the IO environment also introduces new opportunities and challenges for risk assessment and risk management, which will be explored in this book.

IO solutions have the potential to improve risk assessment and management. This may be achieved through (for example) an improved knowledge base for carrying out assessments, access to interdisciplinary expert knowledge, improved presentation of risk-related information and access to updated information. New ways of doing work implies that new inputs for assessments must be identified – either by asking new questions when applying traditional methods or by developing new methods suited to an IO context. Although recent blowouts in the industry indicate that inadequacies in IO solutions have contributed to incidents, adequate IO solutions may have contributed to stopping the progress of events.

References

Albrechtsen, E. and Weltzien, A. 2012. IO concepts as contributing factors to major accidents and enablers for resilience-based major accident prevention, in *Integrated Operations in the Oil and Gas Industry: Sustainability and Capability Development*, edited by T. Rosendahl and V. Hepsø. IGI Global, 1–19.

Chief Counsel's Report. 2011. Macondo – The Gulf Oil Disaster.

Henderson, J., Hepsø, V. and Mydland, Ø. 2012. What is a capability platform approach to integrated operations? An introduction to key concepts, in *Integrated Operations in the Oil and Gas Industry: Sustainability and Capability Development*, edited by T. Rosendahl and V. Hepsø. IGI Global, 353–369.

Herbert, M., Aurlien, J. and Reagan, J. 2008. *ConocoPhillips Onshore Drilling Center in Norway – A Virtual Tour of the Centre Including a Link Up With Offshore.* SPE Paper 11137 presented at SPE Intelligent Energy Conference and Exhibition, Amsterdam, the Netherlands, 25–27 February, 2008.

Johnsen, S.O. 2008. Mitigating Accidents in Oil and Gas Production Facilities, in *Critical Infrastructure Protection II*, edited by M. Papa and S. Shenoi. Springer, 157–172.

Johnsen, S.O., Ask, R. and Røisli, R. 2008. Reducing Risk in Oil and Gas Production Operations, in *Critical Infrastructure Protection*, edited by E. Goetz and S. Shenoi. Springer, 83–98.

Larsen, S., Bjørkevoll, K.S., Gibson, A.K., Gunnerud, V., Lien, D.O., Thorvik, K. and Nystad, A. 2012. *A Stack Model and Capabilities Approach to Investigate Integrated Operations Across Different Industrial Sectors – O&G Industry versus Aviation, Military and Medicine.* SPE Paper 150431 presented at SPE Intelligent Energy Conference and Exhibition, Utrech, the Netherlands, 27–29 March, 2012.

Lilleng, T. and Sagatun, S.I. 2010. *Integrated Operations Methodology and Value Proposition.* SPE paper 128576 presented at the SPE Intelligent Energy Conference and Exhibition, Utrecht, the Netherlands, 23–25 March, 2010.

Montara Commission of Inquiry. 2010. *Report of the Montara Commission of Inquiry.* Montara Commission of Inquiry. Available at http://www.ret.gov.au/Department/Documents/MIR/Montara-Report.pdf.

OLF (Norwegian Oil Industry Association). 2003. *eDrift på norsk sokkel – det tredje effektiviseringsspranget.* [eOperation at the Norwegian Continental Shelf – the Third Efficiency Leap] OLF Report. Available at http://www.norskoljeoggass.no/PageFiles/14295/030601_eDrift-rapport.pdf.

OLF (Norwegian Oil Industry Association). 2005. *Integrated Work Processes: Future Work Processes on the Norwegian Continental Shelf.* OLF Report. Available at http://tinyurl.com/bgta56x/.

PSA (Petroleum Safety Authority). 2010. *Audit of Statoil's Planning for Well 34/10-C-06A A. Petroleum Safety Authority Norway.* Petroleum Safety Authority. Available at http://tinyurl.com/746h79f.

Ringstad, A.J. and Andersen, K. 2007. *Integrated Operations and the Need for a Balanced Development of People, Technology and Organisation.* Presented at the International Petroleum Technology Conference, Dubai, UAE, 4–6 December, 2007 (SPE 11668).

Rommetveit, R., Bjørkevoll, K.S., Ødegård, S.I., Herbert, M., Halsey, G.W., Kluge, R. and Korsvold, T. 2008. eDrilling used on Ekofisk for Real-Time Drilling Supervision, Simulation, 3D Visualization and Diagnosis, in *Proceedings of the 2008 SPE Intelligent Energy Conference and Exhibition.* Amsterdam: The Society of Petroleum Engineers.

Schiefloe, P.M. and Vikland, K.M. 2005. *Årsaksanalyse etter Snorre A hendelsen 28.11.2004* [Causal Investigation into the Snorre A Incident 28.11.2004]. Stavanger: Statoil.

Skarholt, K., Næsje, P., Hepsø, V. and Bye, A. 2009. Integrated Operations and Leadership: How Virtual Cooperation Influences Leadership Practice, in *Safety, Reliability and Risk Analysis: Theory Methods and Application 1*, edited by Martorell et al. London: Taylor & Francis Group, 821–828.

Skarholt, K. and Torvatn, H. 2010. Trust and Safety Culture: The Importance of Trust Relations in Integrated Operations (IO) in the Norwegian Oil Industry, in *Reliability, Risk and Safety: Theory and Application 2*, edited by R. Bris et al. London: Taylor & Francis Group, 1285–1291.

Statoil. 2010. *Brønnhendelse på Gullfaks C.* [Internal investigation report by Statoil after a well event on Gullfaks]. http://www.statoil.no/.

Tinmannsvik, R.K., Albrechtsen, E., Bråtveit, M., Carlsen, I.M., Fylling, I., Hauge, S., Haugen, S., Hynne, H., Lundteigen, M.A., Moen, B.E., Okstad, E., Onshus, T., Sandvik, P.C. and Øien, K. 2011. *Deepwater Horizon-ulykken: Årsaker, lærepunkter og forbedringstiltak for norsk sokkel.* [The Deepwater Horizon Accident: Causes, Learning and Recommendations for the Norwegian Continental Shelf]. SINTEF report A19148.

Weltzien, A. 2011. *Resilience in Well Operations through Use of Collaboration Technology.* Master's thesis at Norwegian University of Science and Technology.

Zuboff, S. 1988. *In the Age of the Smart Machine: The Future of Work and Power.* Oxford: Heinemann.

Chapter 3
Using Human and Organizational Factors to Handle the Risk of a Major Accident in Integrated Operations

Siri Andersen

This chapter explores some of the human and organizational factors associated with the introduction of Integrated Operations (IO) in the oil and gas industry on the Norwegian Continental Shelf, which may increase the risk of a major accident. Based on interviews with industry professionals, it outlines the major concerns of the sector. These concerns include issues such as lack of knowledge and awareness, poor working arrangements, complex collaboration processes, inadequate communication and information capabilities and poor human relations.

The chapter starts with a short and general introduction to human and organizational factors and their influence on major accidents. It then reviews the results of four field studies that provide data on the human and organizational factors that may impact the risk of a major accident following the introduction of IO. The factors identified from these field studies are then compared and contrasted with the results of a review of the literature on the safety implications of IO. This is followed by a discussion of how safety methods may be used to manage these risk factors in the oil and gas industry. Finally, it demonstrates the implementation of a diagnostic checklist to assess the impact of human and organizational factors on drilling and well operations, following the introduction of IO. The application was successfully tested in an audit and was able to pinpoint important flaws in the new IO solutions.

Introduction

The introduction of IO into a company can provoke significant changes in operational concepts, which in turn has an impact on human and organizational factors. Some of these factors may significantly increase the risk[1] of a major accident.

As IO is still a relatively new concept, the opportunities to study actual incidents are limited. Therefore, the empirical data that forms the basis for the current study was gathered from interviews with professionals in the oil and gas industry and other theoretical and empirical studies of the human and organizational impact of IO (Grøtan and Albrechtsen, 2008, OLF, 2007, Ringstad and Andersen, 2006, Skjerve et al., 2008, Tveiten et al., 2008). The focus here is on a few, specific, socio-technical characteristics of IO, in particular: the use of new information and communication technologies, new work processes, the use of real-time data in geographically separate collaborative facilities, and multidisciplinary/organizational/ company working practices (Norwegian Ministry of Petroleum and Energy, 2003–2004, OLF, 2006, 2008).

Although this chapter focuses on the negative impact of IO on safety, it is important to note that there are benefits. Some of these benefits are described in Chapter 2 by Albrechtsen. Participants in the study highlighted several properties of IO that they believed would have the effect of reducing the risk of major accidents. It should also be noted that factors related to security and the management of the implementation phase of IO are not discussed. Finally, emergency preparedness is only briefly touched upon.

Human and Organizational Factors and Major Accidents

That human and organizational factors play a key role in the occurrence of accidents is a truism of safety science and is a premise of this study. Reason (1997) provided an excellent illustration of the point when he demonstrated how latent conditions (organizational

---

1   Risk is defined here as a 'combination of the probability of an event and the consequences of an event' (ISO 17776: 2000). Major accidents are interpreted as accidents with extensive consequences, that is, one or more of the following criteria are fulfilled: at least five fatalities, material damage exceeding 30 million Norwegian kroner or major environmental damage (Sundet, 1990).

and workplace factors) contribute to organizational accidents; in particular the rare but often catastrophic events that can occur within complex modern technologies. Organizational factors are considered to be the strategic decisions and processes that have an impact on individual workplaces through latent conditions (such as inadequate equipment and training). These latent workplace conditions can combine with human nature to produce unsafe acts during operations (at the sharp-end). In this chapter the terms human and organizational factors are used to denote latent conditions.

In general, organizational factors can also refer to particular aspects of the organization. These include *formal aspects* (for example, goals, strategies, division of work and distribution of authority), *informal aspects* (for example, culture and power), *environmental impacts* (for example, uncertainty and external pressure) and *internal processes* (for example, communication, decision-making and resource allocation) (Jacobsen and Thorsvik, 2007). Factors with a specific focus on organizational safety have also been developed; see for example Reiman and Oedewald (2009) and Almklov et al. (2011).

On the other hand, human factors refer to the individual workplace and operators' interactions (physical or mental) with the system (Wickens et al., 2004). The issues addressed by contemporary human factors include, for example, information presentation, workplace design and social and economic impacts (Salvendy, 2006). The distinction between human and organizational factors is, however not always very clear. While initially human factors focused narrowly on human interactions with physical devices, the scope has broadened to include team working and aspects of work and organizational design (Stanton et al., 2005).

This chapter explores the impact of these two themes on major accidents. The following section draws upon the results of four field studies that were conducted with industry professionals in order to identify the human and organizational factors that may emerge following the introduction of IO.

Field Studies

As IO is still a new concept, the opportunities to study actual incidents are limited. Therefore, this study looked at the results of four independent field studies. In these studies, a total of 37 interviews were carried out with professionals in the Norwegian oil and gas industry. The focus of each of the four investigations is outlined below:

A. This study investigated how knowledge for risk decision-making is generated, and looked at the changes that had resulted from the introduction of IO. Nine interviews were conducted at an oil and gas installation with health, safety and environment (HSE), technical safety, maintenance and operational personnel.

B. This study looked at the application of risk analysis methods in the oil and gas industry. It included an examination of their strengths and weaknesses and of the changes that had resulted from the introduction of IO. Eight interviews were conducted with risk analysis personnel working in four operating companies, three consultancies and a drilling contractor (Weltzien, 2010).

C. This study examined the impact of the introduction of IO on the risk of a major accident from five accident perspectives (Haukebø, 2008). The interview transcripts in the original study were re-interpreted for the purposes of this review to highlight human and organizational factors. Seven interviews were conducted with personnel from three operating companies, a vendor and a regulatory organization.

D. This study investigated how personnel from various expert groups perceive risk and risk management in IO (Espeland, 2010). The report provides interview extracts and general summaries of hazards and vulnerabilities in IO. These were used to extract the human and organizational factors. Thirteen interviews were conducted with participants from the oil and gas industry, regulators, consultants and research organizations.

Although each of the four investigations had a different scope, they all addressed human and organizational factors related to IO and major accidents, which is the focus of the current study.

The material was analysed in detail in order to identify the human and organizational factors (themes) that emerged from the interviews (Thagaard, 2009). The aim was to categorize these themes according to their similarities and differences. Matrices provided an overview of the material and supported the development of themes (Miles and Huberman, 1994). The matrices included interview extracts and their associated theme(s). From these matrices, descriptive categories were created to group similar themes/factors. The results are described in the next section.

Human and organizational factors identified

Table 3.1 summarizes the factors believed to have a negative impact on the risk of a major accident. It also shows which of the four studies are related to which factor, and the number of participants that identified the factor.

Table 3.1 **Human and organizational factors related to the introduction of IO**

Category	Human and Organizational Factor	Number of participants[1]	Study
1	Onshore and offshore staff lack competence and awareness		
1a	Offshore workers lack competence	10	A, B, C, D
1b	Offshore workers lack safety awareness	3	C, D
1c	Onshore workers lack competence	9	A, C, D
1d	Onshore workers lack awareness of offshore processes	9	B, C, D,
1e	Lack of education and training	1	B
2	Inadequate work arrangements and complex collaboration processes		
2a	Vague roles and responsibilities in decision-making and work execution	6	B, C, D
2b	Lack of an overall view of offshore and onshore activities affecting offshore systems	7	A, D
2c	Insufficient offshore personnel to detect and prevent hazards and accidents	7	B, C, D
2d	Large number of actors and disciplines involved makes decision-making and collaboration difficult	9	A, B, C, D
3	Insufficient means of communication and lack of information		
3a	Poor quality data and information	8	A, B, C, D
3b	Poor quality and lack of capacity in communication channels	4	B, C, D
4	Poor inter-human relations		
4a	Poor relationships and lack of understanding between offshore and onshore personnel	8	B, C, D
4b	Goal conflicts between onshore and offshore personnel and/or across organizational borders	4	A, C

[1] Number of participants who mentioned the issue.

Study participants held one of the following roles: industry owner/operator, industry vendor, consultant, regulator or researcher. The number of participants within each role and the factors emphasized by each group were as follows:

- Industry owner/operator: 22 participants focusing on all factors except 1b.
- Industry vendor: 2 participants focusing on 1a, 1c, 2a, 2c, 2d, 3a.
- Consultant: 6 participants focusing on 1a, 1b, 1c, 1d, 2a, 2c, 3a, 4a and 4b.
- Regulator: 4 participants focusing on 1a, 1b, 1c, 1d, 2b, 2d, 4a.
- Researcher: 3 participants focusing on 1a, 1c, 1d and 2d.

These numbers demonstrate a limitation of the study, as the results are clearly influenced by the fact that there were many more participants from the industry than consultants, regulators or researchers. Despite the fact that participants do not represent all categories equally, overall the results reflect the opinion of a broad range of actors in the oil and gas industry, which is important for a comprehensive understanding of the situation. Another limitation is that most participants from industry were employed onshore as staff or in middle management positions. Although some participants were employed by specific field organizations and worked on an offshore/onshore rotation, offshore personnel who did not have a specific planning or management role were not represented.

It should be noted that the human and organizational factors shown in Table 3.1 are not independent. Rather, they overlap and are strongly interconnected. For example, there is a link between poor inter-human relations and a lack of communication, another between complex collaboration processes and lack of information, and a third between poor understanding of the situation and conflicting objectives. Neither does Table 3.1 provide an absolute representation of the relationships between factors. While conflicting objectives may be a *result* of poor inter-human relations they may also be the *cause*, or they can occur in the absence of any relationship between factors. However, the classification given here provides an overview of the human and organizational factors that emerged from the four studies and groups them according to common characteristics.

The list of factors shown in Table 3.1 is limited compared to the range of human and organizational factors given in the section describing Human and Organizational Factors and Major Accidents. The reason for this is that the four field investigations had a broad scope and did not provide a detailed examination of the negative safety impacts of IO. Nevertheless, the list is representative of the opinions of industry professionals regarding the most severe impacts on safety and is an indication of the most important human and organizational factors. Table 3.1 also highlights the mix of factors. Category one describes human factors, category two contains organizational factors, and categories three and four are a mix of both. The formal organizational factors mentioned are limited to formal aspects and internal processes, while other factors such as goals, strategies, power struggles and so on do not feature at all. Human factors are limited to information presentation, communication, work design and organization, while other factors such as human characteristics, workplace design, environment and so on are not mentioned. Human factors are most closely related to the fields of cognitive ergonomics (perception, information processing and so on) and macro-ergonomics (optimization of socio-technical systems) as opposed to physical ergonomics (anatomical, physiological, physical characteristics and so on) (Salvendy, 2006).

The following sections describe the factors listed in Table 3.1 in detail.

Onshore and offshore staff lack competence and awareness (Category 1)

Several participants mentioned that the introduction of IO may lead to *a lack of competence and knowledge on the offshore installation* (1a). This was thought to be due to inexperienced personnel involved in daily operations on the installation, and operators with specialist skills being replaced by others with only general or multidisciplinary knowledge (experts and specialists being relocated onshore). Other factors included the proliferation of automated processes and the transfer of decision-making to onshore personnel, which prevented offshore operators from familiarizing themselves with these processes. Participants were concerned that the lack of offshore competence could impair decision-making, safety levels and the ability to carry out repairs. Lack of offshore competence and knowledge was seen as particularly critical during periods when the necessary competence could not be provided onshore (for example, weekends and overnight). Related to this was a perceived *lack of safety awareness among offshore workers* (1b). Participants saw this as being caused by reduced levels of responsibility for offshore safety, insufficient staffing at the installation and a poor understanding of risk due to the inability to shut down the installation.

Lack of onshore competence and knowledge (1c); participants explained that the fact that experts were distributed across various locations made it difficult to ensure necessary general plant knowledge and obtain an overview, for example, of the impact that components might have on each other. Furthermore, participants highlighted the fact that the complexity of the system made it necessary to rely on a wide range of onshore specialists, with theoretical rather than practical knowledge and skills, while inexperienced personnel ran the plant. A closely related factor was that the physical distance to the offshore installation resulted in a *lack of awareness of offshore processes among onshore workers* (1d). Onshore operators were said to lack direct sensory experience and the tacit knowledge and impressions gained while working on the installation, which resulted in a lack of awareness and knowledge. Other points mentioned included the very different work contexts (shifts, surroundings and so on) between offshore and onshore, and the limited ability of onshore workers to know what was happening offshore and gather information. The physical distance to the installation was said to lead to poor or biased decisions due to excess faith placed in second-hand information and lack of direct experience.

Another important issue, mentioned by only one of the participants, underlies all of the factors mentioned in category 1. This is a *lack of education and training* (1e) in the use and functionality of new technology. An attitude to education and training that does not acknowledge the overarching criticality of competence and awareness both onshore and offshore is a decisive factor in whether the other factors mentioned in this category manifest themselves and their actual impact.

Inadequate working practices and complex collaboration processes (Category 2)

Several participants mentioned the *lack of clarity in the roles and responsibilities for decision-making and execution of work* (2a). Two specific examples were given: uncertainty over who was responsible for deciding the severity of an offshore incident, and problems in handing over responsibility in an emergency due to differences between the IO organization and emergency preparedness team members. Another important issue mentioned here was fatigue due to high physical and mental workloads.

A second factor highlighted was the *lack of an overall view of offshore and onshore activities affecting offshore systems* (2b). This was believed to be due to the fact that the many actors involved in the planning and execution of activities were geographically distributed (across disciplines and companies) and that the interfaces between them were poorly managed. Participants thought that information and experience needed to be shared in these interfaces in order to know what others were doing and how that affected one's own activities. A concrete example of this was the fact that onshore personnel had remote access to, and were able to make changes in offshore systems (for example, control and safety systems). Such access made it unclear who had overall control of activities and system status.

A specific factor regarding inadequate working arrangements was that *overall manning levels offshore were insufficient to detect and prevent hazards and accidents* (2c). This manifested as a difficulty in making timely observations (physically and through monitoring systems) and in reacting to unwanted situations and equipment faults. A related concern was there were insufficient offshore personnel with the right knowledge to handle ad-hoc repairs and incidents in a timely manner.

The fourth factor mentioned in this category was that the *large number of actors and disciplines involved made decision-making and collaboration difficult* (2d). Firstly, participants mentioned that the many actors and perspectives involved resulted in an increased diversity of views and opinions. Consequently, it became difficult have an overall understanding of decision-making processes and to control them. Secondly, the number of people participating in the process was believed to result in information overflow, with too much advice to follow, and uncertain command lines (both in normal operation and in emergencies) leading to slow or poor decision-making. Furthermore, some participants mentioned that it could be difficult to know when a decision had actually been taken. They also said that decisions were sometimes made that were not fully understood by those working on the installation. Finally, participants mentioned that the large number of actors involved led to misunderstandings and difficult collaboration due to a failure to utilize competence (the right people, at the right time and in the right place) in a geographically extended environment and in the interface between suppliers and operator.

Insufficient means of communication and lack of information (Category 3)

This factor was closely related to the potential of technology to provide solutions. Participants mentioned that *poor quality data and information* (3a) could be a problem, for example, the loss of, the lack of, and the poor quality of data. They highlighted the insufficient transfer of necessary data (due to a failure of Information and Communication Technology (ICT) infrastructure, for example) and differences in the way the data was represented between personnel from different disciplines. At the same time, they believed that IO could provide too much information, with onshore operators becoming overloaded. They also mentioned that data and information that was transferred to a remote location was not comparable to direct sensory experience and that user interfaces were not always suitable for the integration of new or more technology and functionality.

A related factor was *poor quality and lack of capacity in communication channels* (3b), which also included loss of communication lines. Participants mentioned that this could lead to misunderstandings over when a decision was made, what it consisted of and who was responsible. They were also concerned that face-to-face communication was being replaced by remote communication lines.

Poor inter-human relations (Category 4)

Participants explained that IO could be the cause of *poor relationships and lack of understanding between offshore and onshore personnel* (4a). The concrete examples they gave were mistrust and the loss of collegial relationships between staff members (for example, informal meeting arenas), poor relationships and lack of understanding between onshore and offshore personnel, and onshore personnel lacking knowledge and understanding of offshore work. Furthermore, they pointed to offshore teams not trusting information and guidance provided by onshore support personnel, the difficulty of understanding the conditions experienced by offshore colleagues (stress, emotions and so on) and the many specialist disciplines not being able to understand each other. Some participants mentioned that human relations problems resulted in lack of, or poor communication and collaboration both onshore/onshore, offshore/onshore, between disciplines and vendor/operator.

Another factor mentioned was *goal conflicts between onshore and offshore personnel and/or across organizational borders* (4b) due to differences in opinion on the prioritization of productivity, efficiency and safety. Participants also mentioned that onshore personnel and experts were less concerned about safety than their offshore colleagues. They also highlighted the different understanding of risk and situational awareness among actors who were geographically distributed and from different disciplines, all of which impacted how activities and goals were prioritized. One participant mentioned that not being physically offshore gave a reduced perception of risk.

Other Studies of Human and Organizational Factors in Integrated Operations

While the human and organizational factors described so far are the result of recent interviews with industry professionals, other authors have identified safety implications of IO which to some extent highlights human and organizational factors, for example, Ringstad and Andersen (2006), OLF (2007), Tveiten et al. (2008), Grøtan and Albrechtsen (2008), and Skjerve et al. (2008). Of these five studies, two (Grøtan and Albrechtsen, 2008 and Skjerve et al., 2008), included a review of the other three. Therefore, only these two are referred to in this chapter. The results of these reviews were compared with the empirical material discussed in the previous section in order to include any major points that may not have been identified in the interviews.

Grøtan and Albrechtsen (2008) provide an overview of anticipated organizational changes due to the introduction of IO and discuss the potential for it to have both negative and positive impacts on major accident risk using five different frameworks for organizational accidents. Negative impacts are discussed in terms of 'changes in decision-making processes, communication and collaboration patterns' and 'changes in social and cognitive premises for safe operation and emergency handling'. While most of the negative impacts described in their report correspond to the factors that emerged from the empirical study, they also provide a broader illustration of potential negative impacts when factors are combined. One example is the difficulty of developing an overview of ongoing work across collaboration centres that are geographically distributed also makes it difficult to develop informal procedures, working practices and to coordinate work. Another example is that the ability to improvise is lost when manning levels are reduced, local knowledge is lost and teams who don't know each other have to work together. Therefore, a further issue, which complements the factors discussed in the previous section, is:

> *Communication and information difficulties.* Problems can arise when communication is no longer face-to-face, but instead happens over a remote ICT network. Examples include: a lack of preparedness to handle loss of communication with onshore personnel, and the potential for subtle information (such as disquiet that something is not working properly) to be overlooked. Both of these points add to the explanation of factor 3b.

While Skjerve et al. (2008) discussed in detail the challenges of emergency handling in IO, they also investigated the impact of IO on the need for new or changed Defined Situations of Hazard and Accidents (DSHA). Their review of reports and papers on IO in the Norwegian petroleum industry found several factors that influenced risk, which included: ICT infrastructure; training, teamwork, collaboration technology, and work organization; and maintenance. This study did not reveal any new factors but three important issues that complement the explanation of human and organizational factors in the previous section are:

5. *Poor maintenance due to the long-term outsourcing of maintenance activities.* The review found that it was already challenging to achieve adequate maintenance in situations which went beyond normal or planned activities. There was a danger that further outsourcing would create a situation where it would be difficult to ensure sufficient competence in unusual situations. Moreover, maintenance activities were vulnerable to staff turn-over and it was difficult to allocate responsibility in deviant situations. This is an extension of factor 2a.

6. *Tension and conflict between parties involved in decision-making processes* due to differing interpretations of the information provided by the system. Such tensions and conflicts may eventually lead to a breakdown in the understanding of the situation. Another point is that it is difficult to arrive at an accurate understanding of the situation in a virtual team environment. This issue complements factor 3a as it highlights the importance of the role played by ICT-mediated data in facilitating a shared situational awareness between actors in the IO environment.

7. *Increased automation of normal operations.* This is said to encourage monitoring and discourage intervention. It can also reduce situational awareness and the ability to correctly handle deviations and incidents should they occur. This issue complements factor 1a which addressed automated processes and loss of competence; in addition it points to the impact of automation on the ability of humans to handle deviations and incidents.

Overall, the review of related reports did not reveal any fundamentally new factors in addition to those that emerged from interviews; however they did illustrate and complement other aspects of the problems already identified. The rest of this chapter looks at the implications of these human and organizational factors for safety methods used in the industry and demonstrates how they can be implemented in a diagnostic checklist to assess the impact of the implementation of IO in drilling and well operations.

Human and Organizational Factors and Safety Methods

The human and organizational factors that have been identified here are latent conditions. They may exist in an organization for many years before they combine with local circumstances and active failures (errors and violations committed by operational personnel) to penetrate the system's defences, resulting finally in an accident (Reason, 1997). It is unlikely that any of them would be identified as an immediate or direct cause of an accident and their actual effect depends on how well they are managed during planning, implementation and operational phases of work.

The Norwegian oil and gas industry implements several safety methods in the design and operational phases. In general, these methods have been shaped by the fact that traditionally oil and gas production has been almost solely a matter for the offshore organization. However, many of the factors identified in this chapter manifest themselves in the interface between onshore and offshore organizations, and between vendor and operator organizations. It is therefore important, in safety terms, to consider the entire, geographically distributed socio-technical system as equally important. The safety methods that are applied must be extended to include onshore aspects of the system. The following sections explore some ways in which the factors discussed in this chapter can be incorporated into safety methods.

Extension of offshore safety methods

Many of the human and organizational factors identified in this study are already recognized as important safety preconditions at offshore installations and safety methods have been deployed that take them into account during design and modification phases, especially in offshore control rooms (see for example, NORSOK Standard S-002 2004). For example:

- Function and task analyses ensure that operator task requirements have been identified and system performance has been taken into account. These analyses also form the basis for sound human–machine interaction design.

- Analyses of tasks and the organization of work provide input to the design and organization of working practices, including communications requirements, operating procedures, training, information and control.
- Organization and manning studies describe the necessary skills, experience, responsibilities and operations and maintenance tasks for the various categories of personnel.

However, these efforts are focused on the offshore organization while in IO, traditional offshore functions are spread across geographical, organizational and company borders. This implies that the analyses must be modified to include relevant control and collaboration centres both onshore and offshore, and in operator and vendor organizations.

Extension of human factors safety methods

Within the human factors tradition several methods and techniques have been developed to assess human and organizational factors similar to those discussed here. Two techniques described by Stanton et al. (2005) are of particular interest:

- *Situation awareness assessment methods.* These can be used either to determine the levels of situational awareness offered by new technology and design or to assess situational awareness in operational systems. For example, they can evaluate the impact of new technologies and training interventions on situational awareness, examine the factors that affect situational awareness and evaluate the effectiveness of processes and strategies for acquiring situational awareness.
- *Team performance analysis methods.* These have been developed as a response to the increased use of teams (groups of actors working collaboratively who may be dispersed across a number of different geographical locations) within complex systems. These methods are used in the development of team training interventions, for the evaluation of team

performance, to create task descriptions distributed across a team and to identify required skills, knowledge and abilities.

A stronger focus in the oil and gas industry on these human factors' safety methods may be a useful way to mitigate the impact of the factors identified earlier – such as vague roles and responsibilities, low manning levels, poor quality data and lack of situational awareness, competence and training.

Extension of organizational development methods

For many oil fields, the implementation of IO means that some sort of organizational change must happen in order to incorporate the new concepts. Some responses may therefore be found in Organizational Development models (OD) designed to identify important organizational factors. For example, Weisbord (1983) studied organizations in terms of six topics (frames). This study categorized the problems experienced by organizations according to topic and looked for relations between topics; the end result was an organizational diagnosis that was also able to suggest improvements. A similar approach is the 'pentagon model' (Almklov et al., 2011, Schiefloe and Vikland, 2007) which specifically addresses safety and is therefore particularly relevant. This model takes the view that safety is a function of the interaction between the human, technological and organizational dimensions of an organization. The organization is divided into five topics: structure, technology, relations, interactions and culture. The model can identify problems within and between topics and provides an analysis of both formal and informal aspects of the organization.

Other OD methods that have proven valuable in Norwegian organizations build on the tradition of 'action research', where staff involvement and collective learning are central elements (Levin and Klev, 2002). These methods consider it fundamental that employees' competence and skills are utilized in bringing about change. Change usually requires that personnel are trained and when they are involved in the development process they acquire the new knowledge necessary for working in the new organization. Two examples of methods that have been used for developing good organizational solutions that are accepted by staff are: *search conferences* used for the planning and development of changes through the involvement of employees; and *team development* which seeks to find good working practices, agree on goals and tasks, utilize competence and promote personal development (ibid.). The use of these organizational models and methods may help to mitigate the potential negative effects of the organizational factors previously identified.

Implementation of factors in risk analyses

The Skjerve et al. (2008) study mentioned earlier established that IO is unlikely to have an impact on DSHAs used in the industry (with the possible exception of a few ICT-related DSHAs). Currently, DSHAs are developed as top events (offshore events that imply or may directly lead to fatalities, damage and so on). Top events are also known as hazardous or accidental events. They refer to the event found in the middle of a bow-tie model, where casual events are found on the left-hand side and consequential events on the right-hand side (Rausand and Utne, 2009). However, the report by Skjerve et al. (2008) also concluded that there are many factors that influence safety, which may become relevant with the introduction of IO. The authors point out that these factors must be taken into account in an analysis of events related to current DSHAs and event management.

In fact, the human and organizational factors identified in this chapter should be addressed in the processes that precede DSHAs and in event management, for example, quantitative risk analyses (QRA). They should also be taken into account when performing more limited qualitative design analysis and operative risk analyses, for example, during project modifications or in daily offshore tasks. However, the current approach to QRAs taken by the oil and gas industry has a strong tendency to focus on technical design elements (Andersen and Mostue, 2012). The inclusion of human and organizational factors in risk analyses has been commented on by several authors. While Mohaghegh et al. (2009) suggest a technique to incorporate organizational factors into the probabilistic risk analysis (PRA), Le Coze (2005) discusses the difficulty of introducing the complex nature of organizations into integrated methodologies. Moreover, although there have been several attempts to incorporate organizational factors into quantitative risk analysis, for example, WPAM, MANAGER, MACHINE, IRISK, BORA, these organizational risk analyses are not applied to any great extent in the oil and gas industry (Andersen and Mostue, 2012). If the goal is to be able to decide in advance whether IO will have a negative impact on major accident risk, factors such as those that have been identified in this chapter must be included in risk analyses. One solution could be to utilize methods developed to incorporate human and organizational factors (for example, influence modelling or systemic methods) or to assess these factors separately using human and organizational risk analyses.

This section has given a brief and general discussion of how current safety methods can be used to address the problems raised by human and organizational factors following the introduction of IO. The next section provides a concrete example of how an assessment of the impact of human and organizational factors can be implemented in a diagnostic checklist.

A Diagnostic Checklist for Human and Organizational Factors in an Integrated Operations Context

While the human and organizational factors already described are based on the general characteristics of IO, the choice of IO solutions varies among companies in the oil and gas industry. An analysis of human and organizational factors must therefore always be adapted to the scope of the project and the actual IO solution implemented. To illustrate the point, a diagnostic checklist was developed to investigate the implementation of IO in a company involved in drilling and well operations. The checklist is shown in Table 3.2. This checklist is a condensed version of an actual checklist used by the Petroleum Safety Authority Norway (PSA) in an audit[2] of new technology and work processes in the drilling department of an oil and gas operator.

---

2   The PSA uses a wide definition of audit that includes all contacts between the regulator and the regulated. In other words anything that provides the PSA with the necessary basis to determine whether companies are meeting their responsibility to operate acceptably in all phases of work. This includes

The list of diagnostic questions shown in Table 3.2 was developed in the following steps: 1) a literature review of drilling and well solutions in an IO context; 2) the preparation of some initial questions in a checklist; 3) an audit where the initial checklist was tested; 4) an evaluation of the suitability of the initial questions; 5) comparison of the initial questions with the general factors described in Table 3.1; and 6) the finalization of the condensed list of diagnostic questions shown in Table 3.2.

The literature review[3] included general material on IO concepts and solutions related to drilling and well activities in the oil and gas industry. Based on the literature review, an audit checklist was developed in a joint collaboration of safety researchers and the PSA. The checklist was comprehensive, covering various aspects of IO-related technologies and work processes, including information security issues. This checklist was used in an audit performed by the PSA. The audited company had implemented an onshore support centre, which was continuously manned by two members of staff. This centre only carried out monitoring activities; there was no capacity for remote control. In addition, a larger onshore team was located close to the support centre; this team had responsibility for planning drilling activities and daily support. Furthermore, all offshore data, including the real-time transfer of drilling data, was available onshore. Vendors were represented in the operator's offices and other vendor centres were not used. At this particular installation the introduction of IO had not had much of an impact on manning levels and offshore competence and the offshore drilling crew was still responsible for running drilling operations. Therefore, the human and organizational factors listed in Table 3.1 related to 'lack of offshore competence and awareness' and 'insufficient personnel offshore' are less relevant and given a lower priority in the checklist in Table 3.2.

The audit consisted of a group session with company representatives familiar with the IO solutions that had been

---

supervisory activities, investigations, processing consent applications and conducting meetings with the industry. The audit referred to here is the latter type, that is, a meeting where the company presented their IO practices according to the list of questions described in the text.
3   Published as a confidential SINTEF project memo.

implemented. The company was provided with the main topics in advance and the session was organized into a series of questions from the diagnostic team that were answered by company representatives through presentations and discussions.

Following the audit, the questions in the PSA-checklist were evaluated in collaboration with the PSA. The evaluation resulted in only minor changes to the checklist, the main topics and questions remained unchanged as the evaluating team felt that the questions covered all relevant issues related to the implementation and application of IO-based solutions in drilling and well operations. The checklist has since been deployed in other audits of new technologies and work processes. In general, this exercise demonstrated that the PSA-checklist was capable of characterizing the state of IO implementation in these companies. It was also able to reveal important flaws in the solutions implemented, in terms of human and organizational factors.

This exercise also demonstrated that it was critical to tailor the checklist to the needs of each company. There are significant variations between the IO solutions that have been implemented in the oil and gas industry; they range from approaches that resemble traditional organizations, to continuous collaboration, to the transfer of offshore personnel onshore. The checklist of questions had to be selected and tailored to the solution that had been implemented. The exercise showed that theme used least in the checklist was 'automation'; this is basically because automation is still little used in current drilling solutions.

Finally, to arrive at the list of questions shown in Table 3.2, the PSA-checklist was compared to the general factors described in Table 3.1. The main themes shown in the checklist were arrived at from the topics that emerged from the PSA-checklist rather than the general factors shown in Table 3.1. The purpose of this was to relate the questions to a context familiar to the company. The column labelled 'corresponding factor in Table 3.1' shows the link between the specific, drilling-related questions in the checklist and the general, human and organizational factors shown in Table 3.1.

Table 3.2   Diagnosis checklist for implementation of IO at a drilling installation (condensed list with examples of questions)

Main theme	Main question	Detailed part question	Corresponding factor in Table 31
Onshore and offshore roles	How are the roles played by the onshore centre and offshore personnel currently structured?	Is the composition of onshore and offshore teams adequate and sufficient? How do you ensure that personnel in onshore centres have necessary offshore knowledge? How is knowledge and experience of rig/drilling activities ensured and used onshore? Do onshore teams need to have direct sensory experience to carry out their tasks, and how is this handled? How are differences in onshore and offshore shift hours handled (e.g. handovers)? What is the impact on offshore competence of moving tasks onshore? How is this handled? How is training in new/changed technology and tasks identified and carried out? What role do contractors play in onshore centres (e.g. transfer of experience, sharing of information, communication)? How is work that involves different disciplines carried out (e.g. interdisciplinary teams)? How is communication technology and physical and virtual meetings used to support collaboration? How is responsibility for planning coordinated between distributed actors? What role do onshore centres play in emergency management? What impact has reduced levels of offshore manning had on the ability to handle tasks/incidents?	1a/b/c/d/e 2a/b/c/d

Main theme	Main question	Detailed part question	Corresponding factor in Table 31
Onshore and offshore decision-making	How are the challenges of decision-making that is based on real-time data, collaboration with several actors, access to experts, and so on identified and handled?	What decisions are taken by onshore centres and how are they made? Is the onshore centre's responsibility and authority clearly mediated and understood? Is offshore responsibility and authority clearly mediated and understood? Are those with the authority to take decisions and lines of command clearly established and understood? What are the implications (if any) of using onshore support for offshore decision-making? How are professional authority (expert knowledge, onshore) and the authority to take decisions (offshore) separated? How is the responsibility for safety distributed across onshore and offshore teams? Has it changed with the implementation of IO-solutions?	2a/d 1b
Management of onshore/ offshore interfaces	How are interactions between onshore and offshore actors involved in drilling and well operations organized?	What are the challenges of real-time collaboration between onshore and offshore teams? Are the communication systems/ methods used by geographically distributed actors during operation of good enough quality and sufficiently available? Are there any plans for handling failures in communications systems between distributed actors (for example, onshore/offshore)? How are relationships created and trust established between onshore support teams and offshore decision-makers? How are physical and virtual meetings used to create a common language and understanding between onshore and offshore teams? How are goal conflicts identified and handled? How are differing opinions on operational risk identified and discussed?	2d 3b 4a/b

Main theme	Main question	Detailed part question	Corresponding factor in Table 31
Automation	How does automation support operations?	What are the challenges of automation? What are the criteria for replacing manual processes with automation? How are decisions taken on where and when human intervention may be needed? How is knowledge of automated systems/processes guaranteed?	1a
Collection, handling and use of real-time data	How is the quality of data collection and handling ensured, in terms of: confidentiality integrity availability	Is the correct data collected? How do you ensure that data is available at the right time and there are no delays? How is real-time data integrity ensured? Which safety critical functions/ equipment require good data quality? How are downhole and topside sensors calibrated and classified? How is loss of barriers/well control handled if communication with onshore is unstable or interrupted?	3a 3b
	Are responsibilities and interfaces for the collection of real-time data clearly defined?	Who are the owners of sensors and who is responsible for their maintenance? When is responsibility transferred from one role to another? Are there any vendor requirements regarding data collection and quality?	2a 3a
Simulation and visualization for decision support	How is visualization and simulation used to support operational decision-making?	How is data from different sources combined in decision-making? What were the criteria for choosing to use visualization as a support for decision-making? How is visibility of critical information ensured? How does the user know that visualized information is reliable? How are systems based on real-time data (visualizations, alarms and so on) made user friendly?	3a

Conclusion

IOs generate several challenges in human and organizational factors. The results of the study show that there is no quick fix or one solution fits all to meet these challenges posed by human and organizational factors in the organization. The relative importance of factors will vary depending on the actual IO solution implemented by the company. This chapter has highlighted some safety methods that may be useful in managing and mitigating their potential negative impact. These methods include, for example, a greater focus on human and organizational factors in existing design and modification safety methods, the extended use of human factors safety methods, the extended use of OD methods and the inclusion of human and organizational factors in risk analyses.

A list of diagnostic questions is shown in this chapter to demonstrate how the identification of general human and organizational factors could be applied to the implementation of IO solutions in drilling and well operations. Application of this list of questions in an audit showed that the checklist was able to contribute to safety by revealing important flaws in the IO solutions that had been implemented.

Audits and safety methods are all diagnostic safety management tools. The list of diagnostic questions shown in Table 3.2 could thus be generalized and used in other safety methods, for example, operative risk analyses and OD methods. However, it must not be forgotten that such checklists must always be used with care, and particular attention must be paid to the process of tailoring them to the specific company and solution in question.

References

Almklov, P., Antonsen, S. and Fenstad, J. 2011. *NPM, kritiske infrastrukturer og sammfunnsikkerhet* [NPM, Critical Infrastructures, Public Sector Reorganization and Societal Safety], Report No. 2011. Trondheim, Norway: NTNU. Available at http://tinyurl.com/c5bujlv [accessed 16 July 2012].

Andersen, S. and Mostue, B.A. 2012. Risk analysis and risk management approaches applied to the petroleum industry and their applicability to IO concepts. *Safety Science*, 50 (10), 2010–2019.

Espeland, T.J. 2010. *Perception of Risk in the Environment of Integrated Operations: A Qualitative Study of Four Expert Groups' Understanding of Risk in the Petroleum Sector at the Norwegian Continental Shelf*, Master's thesis. Trondheim, Norway: NTNU. Available at http://tinyurl.com/bsemxqs [accessed 16 July 2012].

Grøtan, T.O. and Albrechtsen, E. 2008. *Risikokartlegging og analyse av Integrerte Operasjoner (IO) med fokus på å synliggjøre kritiske MTO aspekter* [Risk Mapping and Analysis of Integrated Operations (IO) with a Focus on Critical MTO Aspects], Report No. SINTEF 7085. Trondheim: SINTEF. Available at http://tinyurl.com/bnh6hmy [accessed 16 July 2012].

Haukebø, H.M. 2008. *Robuste organisasjoner i integrerte operasjoner* [Robust Organizations in Integrated Operations], Master's thesis. Trondheim: NTNU.

ISO17776:2000, 2000. *Petroleum and Natural Gas Industries – Offshore Production Installations – Guidelines on Tools and Techniques for Hazard Identification and Risk Assessment*, ISO 17776.

Jacobsen, D.I. and Thorsvik, J. 2007. *Hvordan organisasjoner fungerer* [How Organizations Work]. Bergen: Fagbokforl.

Le Coze, J.C. 2005. Are organisations too complex to be integrated in technical risk assessment and current safety auditing? *Safety Science*, 43(8), 613–638.

Levin, M. and Klev, R. 2002. *Forandring som praksis: læring og utvikling i organisasjoner* [Change as Practice: Learning and Development in Organizations]. Bergen: Fagbokforl.

Miles, M.B. and Huberman, A.M. 1994. *Qualitative Data Analysis: An Expanded Sourcebook*. Thousand Oaks, CA: Sage.

Mohaghegh, Z., Kazemi, R. and Mosleh, A. 2009. Incorporating organizational factors into Probabilistic Risk Assessment (PRA) of complex socio-technical systems: A hybrid technique formalization. *Reliability Engineering and System Safety*, 94(5), 1000–1018.

NORSOK Standard. 2004. NORSOK Standard S-002: Working Environment. Oslo: Standards, Norway.

Norwegian Ministry of Petroleum and Energy, 2003–2004. Stortingsmelding nr. 38. Om petroleumsvirksomheten [Report number 38 (2003-2004) to the Storting]. Available at http://tinyurl.com/c5sw668 [accessed 16 July 2012].

OLF (Norwegian Oil Industry Association). 2006. *Verdipotensialet for Integrerte Operasjoner på Norsk Sokkel* [Potential Returns of Integrated Operations on the Norwegian Continental Shelf]. Available at http://tinyurl.com/ c7br99m [accessed 16 July 2012].

OLF (Norwegian Oil Industry Association). 2007. *HMS og Integrerte Operasjoner: forbedringsmuligheter og nødvendige tiltak* [HSE and Integrated Operations: Improvement Opportunities and Necessary Actions]. Available at http://tinyurl.com/bsdc5j9 [accessed 16 July 2012].

OLF (Norwegian Oil Industry Association). 2008. *Integrated Operations in New Projects*. Available at http://tinyurl.com/cemukq5 [accessed 16 July 2012].

Rausand, M. and Utne, I.B. 2009. *Risikoanalyse: teori og metoder* [Risk Analysis: Theory and Methods]. Trondheim: Tapir akademisk forl.

Reason, J. 1997. *Managing the Risks of Organizational Accidents*. Aldershot: Ashgate.

Reiman, T. and Oedewald, P. 2009. *Evaluating Safety-critical Organizations – Emphasis on the Nuclear Industry*, Report No. 2009-12. VTT, Technical Research Centre of Finland. Available at http://www.vtt.fi/inf/julkaisut/ muut/2009/SSM-Rapport-2009-12.pdf [accessed 1 May 2012].

Ringstad, A.J. and Andersen, K. 2006. *Integrated Operations and HSE – Major Issues and Strategies*. Presented at the SPE International Conference on Health, Safety and Environment in Oil and Gas Exploration and Production, Abu Dhabi, UAE, 2–4 April 2006 (SPE 98530).

Salvendy, G. 2006. *Handbook of Human Factors and Ergonomics*. Hoboken, NJ: Wiley.

Schiefloe, P.M. and Vikland, K.M. 2007. Når barrierene svikter. Gassutblåsningen på Snorre A, 28.11.2004 [When barriers fail. The gas blowout on Snorre A, 28/11/2004]. *Søkelys på arbeidslivet*, 24(2), 83–101.

Skjerve, A.B., Albrechtsen, E. and Tveiten, C.K. 2008. *Defined Situations of Hazard and Accident Related to Integrated Operations on the Norwegian Continental Shelf*, Report No. SINTEF A9123. Trondheim, Norway: SINTEF. Available at http://tinyurl.com/7hg825a [accessed 16 July 2012].

Stanton, N.A., Salmon, P.M., Walker, G.H., Baber, C. and Jenkins, D.P. 2005. *Human Factors Methods: A Practical Guide for Engineering and Design*. Aldershot: Ashgate.

Sundet, I. 1990. *Kartlegging av storulykker i Norge* [Mapping of Major Accidents in Norway]. Report No. STF75 A90029. Trondheim: Norwegian Institute of Technology.

Thagaard, T. 2009. *Systematikk og innlevelse: en innføring i kvalitativ metode* [Systematics and Insight: An Introduction to Qualitative Methods]. Bergen: Fagbokforl.

Tveiten, C.K., Lunde-Hanssen, L.S., Grøtan, T.O. and Pehrsen, M. 2008. *Hva innebærer egentlig Integrerte Operasjoner? Fenomenforståelse og generiske elementer med mulige konsekvenser for storulykkespotensialet* [What are Integrated Operations? Understanding the Phenomena and Generic Elements with Potential Consequences for Major Accidents], Report No. SINTEF A7078. Trondheim, Norway: SINTEF. Available at http://tinyurl.com/cmvc53c [accessed 16 July 2012].

Weisbord, M.R. 1983. *Organisationsdiagnos: en handbok med teori och praktiska exempel* [Organizational Diagnosis: A Handbook of Theory and Practical Examples]. Lund: Studentlitt.

Weltzien, A.H. 2010. *Methods for Risk Assessment of Integrated Operations in the Norwegian Petroleum Industry.* Master's thesis. Trondheim: NTNU.

Wickens, C.D., Lee, J., Liu, Y.D. and Gordon-Becker, S. 2004. *An Introduction to Human Factors Engineering.* Upper Saddle River, NJ: Pearson/Prentice Hall.

Chapter 4
Assessing Risks in Systems Operating in Complex and Dynamic Environments

Tor Olav Grøtan

This chapter focuses on an exploration of complexity in relation to the impact of Integrated Operations (IO) on safety. The assumption is that complexity can have a detrimental (pathogenic) impact on safety. However, it also has the potential to develop safety-reinforcing (salutogenic) properties that go beyond the absence of failure. Resilience and complexity are intricately interwoven. Resilience can be seen as an adaptive capacity that is a response to complexity, but that may also contribute to, or leverage it. Resilience as a property of safety may both counter pathogenesis and embody salutogenesis.

A key implication of complexity is that knowledge of past events may not be sufficient to prepare for new, dangerous events and circumstances. Therefore, I propose that in addition to event-based assessments, risk management should analyse the type of complexity under which the system operates. A risk proxy encompassing both pathogenic and salutogenic properties is proposed and a framework for managing the proxy is outlined.

Integrated Operations and Risks

This chapter looks at the relationship between IO, complexity, resilience and risk assessment. It takes as a starting point the assumption that IO may inherently increase system dynamics (SD) and complexity, thereby making future system states more difficult to predict. It also assumes that IO can encourage organizations to deal with known risks in a way that neglects the impact of change in new operational environments. Looked at from this perspective, IO is a potential breeding ground for systemic risks (Renn, 2008) and it is worth investigating the link that might exist between IO and safety, in order to assess whether it is necessarily beneficial.

Integrated operations

IO in the oil and gas industry encompasses various agendas ranging from the deployment of new technology to wide-scale business transformation. For example, Grøtan et al. (2010: 2) portray IO as follows:

> The development towards increased integration is signified by increased bandwidth in the digital infrastructure, standardisation of data, integrated ICT applications and work processes utilising the digital infrastructure, as well as new work processes aimed at achieving improved and more efficient analysis and decision processes. This line of development is strongly linked to [...] more efficient reservoir exploitation, optimization of exploration and operation processes, and ambitions for long-term, managed development of fields and installations (OLF 2003) [...] From being primarily focused on technology development and application, the development of IO now takes new directions:
> * increased focus on challenges related to new work processes, integration of information throughout whole value chains and ICT vulnerability;
> * wide recognition of the prime significance of human and organizational factors for the success of IO;
> * development of 'virtual plants' comprised by onshore and offshore elements that are transformed into a fully integrated functional community.

This description highlights that IO is an evolving concept; its evolution is based on the potential offered by Information and Communication Technology (ICT). A more detailed presentation of IO can be found in Chapter 2 by Albrechtsen.

Integrated Operations and Safety: Beyond the Absence of Failure

IO has safety implications both in terms of threats and new opportunities.

> *IO to prevent unwanted events.* Intuitively, IO might improve safety by preventing or countering well-known failure mechanisms. This is consistent with the traditional approach to safety management that looks at what may go wrong and can take the form of prescribing operational patterns, and discovering and preventing errors, anomalies and deviations. The approach relies on a disease metaphor that corresponds to the concept of pathogenesis in medicine. Reason's

(1997) view that latent conditions contribute to organizational accidents rests on the same metaphor.

IO to foster wanted events. In addition to preventing things from going wrong, IO can also facilitate *salutogenesis*.[1] Resilience Engineering adopts this approach, which focuses on human adaptive capacities to look for what may go right. In this respect, Resilience Engineering sees humans as a resource rather than a problem. In doing so, it distinguishes between Theory W and Theory Z (Hollnagel, 2010). Theory W claims that things go right because of a normative prescription of behaviour and compliance with procedures. Theory Z claims that things go right because of the capacity of humans to recognize problems, adapt to situations, interpret procedures to match the prevailing conditions and contribute to system recovery in degraded conditions. Theory W may be seen as an attempt to preclude pathogenesis, while Theory Z is more like an attempt to stimulate salutogenesis. Although the principles of Resilience Engineering encompass both cases, they clearly embrace the latter through making extensive reference to concepts of variability and emergence, which render the usefulness of the notion of 'human error' rather questionable (Grøtan et al., 2011). The question arises of whether IO can be designed in ways that support both Theory W and Theory Z. Further information about Resilience Engineering as applied to IO can be found in Chapter 14 by Hollnagel.

Large socio-technical systems, where IO is of most interest, are characterized by a large number of diverse combinations of both human and technical components and a certain degree of complexity. In order to understand the pathogenic and salutogenic properties of such systems, one needs to understand the basics of the concept of complexity and its relation to safety. This is the subject of the next section.

System Complexity, Accidents and Risk

In everyday life, 'complex' is often used to emphasize or highlight the idea of 'complicated' or 'difficult', apparently

---

1   The term salutogenesis was coined by Antonovsky (1987, 1990). It refers to 'health-producing' rather than 'disease-producing' (pathogenic) factors. Here, salutogenesis is used as metaphor for the contribution of human and organizational factors to what goes right in demanding and exceptional circumstances. Salutogenesis is thus an indicator of 'organizational health' that goes beyond the mere absence of 'organizational disease'.

without any specific meaning other than the idea that something is far from trivial to resolve and manage. However, in socio-technical systems, complexity is a property that can produce unexpected patterns of behaviour. These patterns of behaviour are due to the number of components, their nature and the changing relationships between them. Because they are difficult to fully predict, complex phenomena (and systems) are sometimes characterized as residing 'at the edge of chaos' (for example, Sawyer, 2005: 3). Although a state of chaos renders any assumptions about risk worthless, it is a state most socio-technical systems are rarely expected to enter – and if they do, it is called an accident or incident. Nevertheless, complexity implies an unexpected proximity to chaos, which may be (dangerously) overlooked by traditional indicators. At the same time, complexity has a salutogenic potential, in that it provides and maintains flexibility and offers favourable conditions for adaptations and transformations resulting from human and organizational actions.

However, complexity does not have a consistent meaning across scientific disciplines. There is a wide diversity of sources and definitions, some of which are listed by Grøtan et al. (2011). Grøtan and Størseth (2011) provide a closer examination of the idea of the complex accident. They draw upon the work of Sawyer (2005) who focuses on *social emergence* dynamics that appear in between the macro and micro system properties that normally receive attention. This approach sees socio-technical complexity as more radical than the complexity found in the material world. This approach to complexity is consistent with Weick's notion of elusive, but crucial, organizational activity that goes on 'behind the rational façades of the impermanent organization' (Weick, 2009). Weick (ibid.) uses the label *distributed sensemaking* to summarize key aspects of complexity. Despite differences in the way complexity is described and defined, its pathogenic and salutogenic properties are clearly maintained.

Complexity must be seen as a main contributor to accidents (pathogenesis), particularly given its capacity to render systems almost impossible to predict (a classical, but not exhaustive discussion of this topic is offered by Perrow, 1984). From this perspective, a complex accident can be seen as an unexpected system state that becomes unmanageable and unstoppable, thereby departing from the intended, safe operational envelope. However unexpected inertia may also be the cause of complex accidents. For example, Grøtan and Størseth (2011) claim that complex accidents can be caused by latent interactions that silently solidify (through social emergence) into system properties that are unable to respond to unexpected conditions and inhibit salutogenesis. Finally, complex accidents can also be characterized by unexpected conditions that combine with trivial circumstances. Extraordinary couplings may then appear that have unexpected properties (ibid.).

Complexity is not an explicit and easily identifiable pattern that can be related to the input/output characteristics of systems. Instead, it is an inherent property and its consequences can take different forms. The remainder of this chapter will focus on the three forms outlined below (Grøtan et al., 2011) and their implications for the validity of risk assessments:

- *Manifest complexity* describes systems that are constantly changing and whose transformations can be understood only in retrospect. The dynamics of social emergence (Sawyer, 2005) may be identified in real-time, but their implications are very uncertain. Although the organization's 'rational façade' (Weick, 2009) is persistently challenged, it remains in place due to the lack of credible alternatives. Manifest complexity is an indication that current risk assessments may be about to become invalid, as key assumptions about system behaviour are rendered uncertain.
- *Latent complexity* describes a situation where only intermittent complexity is seen in a system that is otherwise assumed to be stable. Here, the organization may be tempted to optimize efficiency based on how the system has functioned to date and fail to give sufficient attention to potential deviations and their consequences. In these circumstances, the 'rational façade' is more credible and easier to maintain. Latent complexity appears to have no effect on risk assessments until they suddenly become outdated; moreover their loss of validity may go unnoticed due to the short duration of the

actual change in operating conditions. Latent complexity has also been termed 'wildness-in-wait'.[2]

- *Un-order.* This third form of complexity is based on the premise that an apparently non-complex system is ordered in the sense that it is stable and predictable. However, the notion of un-order (based on Kurtz and Snowden, 2003) implies that neither manifest nor latent complexity necessarily lead to chaos (an accident), and that both may function as junctions or transition phases between different states of order. The concept implies that both the 'rational façade' and risk assessments should be revised at the same time as actual transitions.

As far as safety is concerned, the above suggests that the assumptions underlying risk assessments must be maintained and revised taking into account the potential implications of complexity. Complex systems pose challenges for risk assessment in at least two ways:

- the difficulty of capturing their properties; and
- uncertainty and the social dimensions of knowledge.

These challenges are examined in detail in the following section.

The Challenges of Complexity for Risk Assessment

It makes perfect sense for the risk assessment process to look for similarities between systems. It also makes sense to re-use and modify existing assessments based on the assumption that system properties can be measured by the same standard and are repeatable. Such assumptions are probably valid in similar technological systems but there are challenges that lie beyond

---

2 'Wildness-in-wait' dates from a citation in Bernstein (1996) of Chesterton (1909), 'The real trouble with this world of ours is not that it is an unreasonable world, nor even that it is a reasonable one. The commonest kind of trouble is that it is nearly reasonable, but not quite. Life is not an illogicality; yet it is a trap for logicians. It looks just a little more mathematical and regular than it is; its exactitude is obvious, but its inexactitude is hidden; its wildness lies in wait.'

the technological perspective. This is the issue that this section addresses.

Risk assessors may see IO as offering increased managerial control over the maintenance of a safe operational envelope. However, using the arguments from the previous section, I will assume the opposite. Control can never eliminate the need to be flexible. Indeed, systemic risks (Renn, 2008) may become relevant to an extent not seen before. The key idea put forward here is that IO is a socio-technical system that is inherently complex (Hanseth and Ciborra, 2007). This creates at least two challenges for risk management, outlined below.

Challenge I: Capturing the properties of complex systems

The shortcomings of traditional approaches become apparent when human and/or organizational issues are included in risk assessments of complex systems, unusual situations and rare circumstances (Le Coze, 2005, Grøtan et al., 2011). Traditional, event-based or scenario-based approaches overlook blind spots that are the result of latent complexity. Although the characteristics of a particular accident may of course be of some value to prevent events recurring, it is highly improbable that a complex combination of conditions will occur twice in exactly the same way. When there is latent complexity in the system, a small difference in conditions can make a huge difference to the resulting events and their consequences.

Manifest complexity resists attempts to translate system characteristics into parameters. The current trend of using fictional parameters (Aven, 2009) implies that manifest complexity can be actively ignored or ruled out, due to a (flawed) belief that (fictional) parameters are broadly valid generic abstractions.

The challenge posed by un-order can, in principle, be overcome by frequent updates to risk assessments. However, these updates also require a concomitant review of basic, underlying assumptions that mirror the actual pattern of un-order, to an extent that is not easy to achieve in practice.

Challenge II: Uncertainty and the social dimensions of knowledge

Renn (2008: 194) highlights the limitations of subjective probabilities put forward by experts, 'When ignorance is likely to dominate the picture, or the uncertainties also cover costs and social impact, technical experts are less suited and certainly not legitimized to feed in their probability judgements.'[3] This resonates with the views of Grøtan et al. (2011) who take issue with reliance on experts who use probability concepts to address problems of complexity that are fundamentally indeterminate. Aven (2009) goes even further and claims that both probabilistic assessments and the Bayesian approach are inapplicable due to their overreliance on fictional parameters. The solution put forward by Aven is that the broader concept of uncertainty should be expressed in a probabilistic language, in order to represent the assessor's lack of knowledge (the same approach is described by Vatn in Chapter 13). However, the concern put forward here is that the use of probabilities can hide important issues related to the social aspects of lack of knowledge. Even a team of multidisciplinary experts who combine their respective knowledge would find it difficult to a) find common ground between different scientific disciplines and b) account for the role of the social, cultural and political context in which assessments are carried out.

Bernstein (1996) notes that the issue is that of the extent to which the past determines the future. Therefore, despite the noble intentions of risk assessors and any consensus achieved between them, probabilistic language can carry implicit assumptions that place undue importance on the past. Moreover, it seems questionable to think that it provides an adequate approach to risk communication and dialogue about complex, indeterminate systems. Both Bernstein (ibid.) and Luhmann and Rosa (cited by Johansen, 2010) remind us that the word risk derives from the early Italian word *risicare* (to dare). In this sense, risk does not equal fate. Rather, it is a choice made by the humans that are part of the system and not simply the system designers or risk assessors.

---

3   Note that Renn's terms 'uncertainty' and 'ambiguity' correspond to the term 'complex' used here.

Proxy Criteria as an Alternative Strategy

In response to the dilemma of probabilistic knowledge applied to partly unpredictable systems, Renn (2008: 194) proposes the following,

> If uncertainties cannot be resolved through scientific experiments, modeling or simulation, and subjective probabilities are of little validity, proxy criteria can be used for judging the potential for harm. [...] They have one characteristic in common: they can be assessed scientifically and quantified with a level of precision similar to that of risk assessment.

Examples of proxy criteria given by Renn (ibid.) include ubiquity, bio-accumulation, potential for misuse and high potential for disaster. Complexity must be characterized by a proxy variable that makes reference to system properties rather than simply the potential for harm. In particular, proxy criteria must focus on the impact of both pathogenic complexity and salutogenic resilience that aim to counter complexity. This use of proxy criteria enables a different approach to risk management in complex systems: it supports the assessment of possible future system states that is not based on past events.

Proxy Criteria that Incorporate Complexity

The Cynefin sensemaking framework proposed by Kurtz and Snowden (2003) and shown in Figure 4.1 provides the basic constituents of a risk proxy for complex systems hereafter termed the Complex Risk Proxy (CRP). The four Cynefin domains (Figure 4.1, left-hand side) mirror the distinctions between:

- the accidental breakdown and its outcome (corresponds to 'chaos');
- manifest complexity (corresponds to 'complex');
- latent complexity, reflected by the potential inherent in the ordered domains of 'known' and 'knowable';
- un-order corresponds to the successive, ongoing transitions between (non-chaotic) domains.

The Cynefin-based CRP can represent pathogenic potentials by adverse movements in the Cynefin space (Figure 4.1, right-hand side,

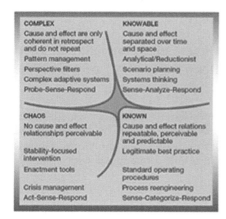
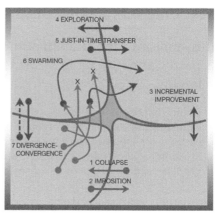

Figure 4.1   **The Cynefin framework (Kurtz and Snowden, 2003)**
                   as a constituent of the Complex Risk Proxy (CRP)

movements 1 and 7[4]). The assessment of various instantiations of salutogenic potentials (related to movements 2, 3, 4, 5, 6 and 7) can in turn relate to very different conceptualizations of resilience (for example, Størseth et al., 2010, Grøtan, 2011, Hollnagel, Chapter 14).

The challenge for the risk assessor is then to understand which mode the system might enter, the potential transition from one mode to another and the possible consequences. Of course, it is still possible to express these assessments in probabilistic language.

Managing Risks via the Complex Risk Proxy

The CRP proposed above requires a managerial framework that encourages attention be paid to non-traditional areas of activity in the socio-technical system; it invites a look 'behind the rational façades' in order to spot signs of social emergence between macro and micro (Sawyer, 2005), or other indicators of complexity. Moreover, it demands that risk management becomes a sensemaking process at institutional level, rather than a 'decision machine' that is fitted into the rational façade. In this context, sensemaking refers to the reality of everyday life that must be seen as an ongoing process (Weick, 2001). It takes shape as individuals attempt to create order and make retrospective sense of the situations they find themselves

---

4   That is, Divergence without succeeding Convergence.

in. It is a process of committed interpretation, where stability is introduced into an equivocal flow of events, through explanations that increase social order (ibid.: 15).

On the basis of this understanding, the use of the CRP in risk assessment can be facilitated by the concept of Organizational Encounters with Risk (OEWR), proposed by Hutter and Power (2005). They see the organization as a critical element with respect to risk as it is where hazards and attendant risks are conceptualized, measured and managed. They look at the 'risks of risk management' and ask how organizations experience the nature and limits of their own capacity to respond to external shocks and disturbances.

Hutter and Power (ibid.) argue that a rational decision theory can be counterproductive. They claim that the traditional separation between the environment and risk management overlooks the role of context, sequence and attention capacity. The organizing process itself is therefore a source of risk. Hutter and Power argue that a new contextualization of risk is needed. The main aspects of OEWR that support the CRP are:

- the *organization of attention*, how organizational priorities determine where and what to look for, what is actually observed (or not) and where attention is paid;
- individual and institutional *sensemaking processes* and the link between them;
- the ability to *re-make the organization* when needed or when sense is lost.

The foundation provided by the OEWR framework means that CRP properties can be addressed, evaluated, interpreted and made sense of in ways that go beyond deviation or failure to comply. The OEWR approach actively supports and requires attention to be given to what is going on behind the façades, to a degree that would escape a traditional, compliance-oriented safety management system. The combination of CRP and OEWR thus promotes a set of complementary forms of safety that resemble those proposed by Grøtan and Størseth (2011), namely:

- *A self-evident* safety based on event-orientation, compliance and deliberate adjustment, in which pathogenesis is resisted *at* the rational façades.
- *An elusive* and emergent safety founded on resilience, in which salutogenesis is facilitated *behind* the very same façades.

Example: The Macondo Accident

The following sample from Hopkins (2011: 1421) illustrates the relevance of the OEWR framework in relation to the recent Macondo disaster:

> About seven hours before the Gulf of Mexico oil well blowout of 2010, a group of four company VIPs helicoptered onto the drilling rig in question, the Deepwater Horizon. They had come on a 'management visibility tour' and were actively touring the rig when disaster struck. [...] There were several indications in the hours before the accident that the well was not sealed and was at risk of blowing out. These indicators were all either missed or misinterpreted by the rig staff. The touring VIPs [...] had all worked as drilling engineers or rig managers in the past and had a detailed knowledge of drilling operations. Had they focused their attention on what was happening with the well, they would almost certainly have recognized the warning signs for what they were, and called a halt to operations. But their attention was focused elsewhere, and an opportunity to avert disaster was lost. [...] A major purpose of the visit was to emphasize the importance of safety, and yet the visitors paid almost no attention to the safety critical activities that were occurring during their visit.

The above extract clearly indicates a possible connection between the accident and a lack of adherence to OEWR principles, despite the substantial impact of IO on operations at the platform:[5]

- *Organization of attention*: senior management were not indifferent to safety, but were focused on occupational, rather than process safety.
- *Sensemaking processes*: they were not able to identify and make sense of the catastrophic (process) safety problem that was developing literally under their feet.
- *Re-making the organization*: they were replicating safety found at other, similar installations, a decision that in retrospect turned out to be unproductive in terms of process safety.

---

5   See Chapter 2 by Albrechtsen for more details about the IO aspects of this operation.

It would clearly be a grotesque exaggeration to claim that the CRP proxy and the OEWR framework would have helped avoid the Macondo accident. However, these observations and considerations suggest that the organization was 'at the edge of chaos' and it could have responded by:

- a sensitivity to complexity issues;
- a more thoughtful attitude towards the organization of attention;
- facilitating sensemaking rather than decision-making processes;
- less reliance on standardized organizational schema; and
- more coherence between individual and institutional sensemaking.

Conclusion

The key arguments proposed in this chapter can be summarized as follows:

- IO refers to a class of socio-technical systems that are so complex that their behaviour is not always predictable. This complexity implies an increased potential for accidents that should be addressed by considering complexity at a conceptual level.
- Socio-technical complexity conveys pathogenic ('disease-creating') as well as salutogenic ('health-creating') potentials. The latter can be seen as the bedrock of resilience.
- These issues create a series of challenges for risk assessment, including reliance on a probabilistic, event-based approach to potentially rare events.
- The alternative solution (Complexity Risk Proxy) consists of an assessment of some system properties in order to understand the states a complex system can enter, and the dynamics between these states.
- The Cynefin sensemaking framework is an example of a CRP that may be useful for orchestrating the necessary updates to risk assessments and for complementing traditional risk management schemes.

- The concept of OEWR may help in the operation and use of the CRP.
- The Macondo accident supports the hypothesis that a CRP/ OEWR approach can help to detect pathogenic conditions or shifts between system modes.

References

Antonovsky, A. 1987. *Unraveling The Mystery of Health – How People Manage Stress and Stay Well*. San Francisco: Jossey-Bass Publishers.

Antonovsky, A. 1990. *Studying health vs. studying disease*. Lecture at the Congress for Clinical Psychology and Psychotherapy, 19 February, Berlin.

Aven, T. 2009. Perspectives on risk in a decision-making context – review and discussion. *Safety Science*, 47, 798–806.

Bernstein, P.L. 1996. *Against the Gods. The Remarkable Story of Risk*. John Wiley and Sons Inc.

Chesterton, G.K. 1909. *Orthodoxy*. New York: Lane Press (Reprinted by Grenwood Press, Westport, 1974).

Grøtan, T.O. 2011. *The Stratified and Dialectical Anatomy of Organizational Resilience*. 4th Symposium on Resilience Engineering, 8–10 June, Sophia-Antipolis, France.

Grøtan, T.O., Albrechtsen, E., Rosness, R. and Bjerkebæk, E. 2010. The influence on organizational accident risk by integrated operations in the petroleum industry. *Safety Science Monitor*, 14, Article 5, 1–11.

Grøtan, T.O. and Størseth, F. 2011. *Integrated Safety Management based on Organizational Resilience*. ESREL 2011, 18–22 September, Troyes, France.

Grøtan T.O., Størseth, F. and Albrechtsen, E. 2011. Scientific foundations of addressing risk in complex and dynamic environments. *Reliability Engineering and System Safety*, 96, 706–712.

Hanseth, O. and Ciborra, C. (eds) 2007. *Risk, Complexity and ICT*. Northampton, MA: Edward Elgar.

Hollnagel, E. (ed.) 2010. *Safer Complex Industrial Environments: A Human Factors Approach*. Boca Raton, FL: CRC Press.

Hopkins, A. (2011). Management walk-arounds: Lessons from the Gulf of Mexico oil well blowout. *Safety Science*, 49, 1421–1425.

Hutter, B. and Power, M (eds) 2005. *Organizational Encounters with Risk*. Cambridge: Cambridge University Press.

Johansen, I.L. 2010. *Foundations of Risk Assessment*. ROSS (NTNU) Report 201002. Available from http://www.ntnu.no/ross.

Kurtz, C.F. and Snowden, D.J. 2003. The new dynamics of strategy: Sense-making in a complex and complicated world. *IBM Systems Journal*, 42, 462–483.

Le Coze, J.C. 2005. Are organisations too complex to be introduced in technical risk assessment and current safety auditing? *Safety Science*, 43, 613–638.

OLF (Norwegian Oil Industry Association). 2003. *eDrift på norsk sokkel – det tredje effektiviseringsspranget*. [eOperation at the Norwegian Continental Shelf – The Third Efficiency Leap] OLF Report. Available from http://www.norskoljeoggass.no/PageFiles/14295/030601_eDrift-rapport.pdf.

Perrow, C. 1984. *Normal Accidents: Living with High-risk Technologies*. New York: Basic Books.

Reason, J. 1997. *Managing the Risks of Organizational Accidents*. Aldershot: Ashgate.

Renn, O. 2008. *Risk Governance*. Earthscan.

Sawyer, R.K. 2005. *Social Emergence. Societies as Complex Systems*. Cambridge: Cambridge University Press.

Størseth, F., Albrechtsen, E. and Eitrheim, M.H.R. 2010. Resilient recovery factors: explorative study. *Safety Science Monitor*, 14(2), 1–16.

Weick, K.E. (2001). *Making Sense of the Organization*. Wiley Blackwell.

Weick, K.E. (2009). *Making Sense of the Organization. Volume Two. The Impermanent Organization*. Wiley.

Chapter 5
Lessons Learned and Recommendations from Section One

Denis Besnard and Eirik Albrechtsen

Take-home Messages from Chapters 2–4

Albrechtsen (Chapter 2) explained what Integrated Operations (IO) is about and its impact on the prevention of major accidents. The main operational messages are as follows:

- Adequate IO-based solutions can improve risk management.
- The advantages of IO include better risk-informed decisions as a result of: an improved and more up-to-date risk assessment knowledge base; access to interdisciplinary expert knowledge; and improvements in the presentation of risk-related information.
- Inadequate IO-based solutions can increase the risk of major accidents. There have been inadequacies in IO-related solutions in recent major incidents in the industry.
- New ways of assessing risk and safety imply the need for new inputs, either by asking new questions when applying traditional methods or by developing new methods suited to an IO context.
- There is a potential transfer of risk management experience both to and from other sectors.

Andersen (Chapter 3) carried out fieldwork in the oil and gas domain and outlined the development of a human factors' checklist for risk assessment. The main operational messages are as follows:

- The introduction of IO generates several challenges in terms of human and organizational factors.
- The chapter describes a checklist that can be used to identify IO-related human and organizational factors, which can be used in various safety methods.
- Although the checklist is not exhaustive, it can integrate data from a wide variety of sources.
- If the checklist is intended to cover organizational aspects of risk management, it must focus on latent conditions.
- The items covered by the checklist must not be restricted to offshore operations; other parties, such as onshore operators and regulators must also be involved.

Grøtan (Chapter 4) addressed complexity in safety-critical systems from a theoretical perspective. The main operational messages are as follows:

- IO refers to a class of socio-technical systems so complex that their behaviour cannot always be predicted.
- Socio-technical complexity creates the potential for both disease as well as health.
- Socio-technical complexity also creates various challenges for risk assessment, including reliance on a probabilistic, event-based approach to potentially rare events.
- The alternative (risk proxy) consists of assessing system properties in order to understand the states a complex system can enter, and the dynamics between these states.

Recommendations

On the basis of these chapters, the following operational recommendations can be made:

- IO-based solutions may offer potential improvements in major accident prevention. However, inadequacies in these solutions are risk influencing factors.
- Careful thought must be given to the development, implementation and application of new technology and work practices. The management of such changes requires the use of various methods for the identification and assessment of safety and risk.
- Something that is easily forgotten is that what you look for is what you find (Lundberg et al., 2009). New work practices and new technology require new questions to be asked about the identification and assessment of hazards and risk.

References

Lundberg, J., Rollenhagen, C. and Hollnagel, E. 2009. What-You-Look-For-Is-What-You-Find – The consequences of underlying accident models. *Safety Science*, 47(10), 1297–1311.

Section Two
Operations and Risk Assessment

Chapter 6

On the Usefulness of Risk Analysis in the Light of Deepwater Horizon and Gullfaks C

Jørn Vatn and Stein Haugen

In this chapter we review the incident on the Gullfaks C platform in the North Sea and the Deepwater Horizon disaster in the Gulf of Mexico. The aim is to investigate the types of risk analysis that were used in the planning of drilling and well operations. Standard practice in the North Sea is to divide risk analysis into three broad categories, characterized as strategic, qualitative design and operative. These categories help to identify the type of risk analysis that could have discovered the problems that contributed to the two events. Based on our findings, we propose several ways to improve the quality of risk analysis.

Introduction

This chapter discusses the usefulness of risk analysis as a tool for decision support. It analyses the findings from two cases, the Deepwater Horizon disaster and the Gullfaks C near-accident and discusses the support that risk analysis can provide in the planning and preparation of operations. In this context 'usefulness' relates to whether risk analysis is able to provide necessary and useful input to decisions affecting the level of risk. The following questions are addressed:

- What types of risk analyses were used in the planning of drilling and well operations and how did these analyses support the decision-making process? The answer to this question provides some background information and sets the scene for further discussion.

- Did the risk analyses provide adequate support for decision-making? Here, the answer draws upon material from the investigation reports. It provides examples of the risk analyses that were carried out and examines how they were used in the decision-making process.
- How can risk analyses and the way the analyses are used be improved to provide better support for decision-making? The answer to this question draws some conclusions from the previous discussions.

Risk analysis is understood to be a structured and documented process where the aim is to identify undesired events and express uncertainty regarding their occurrence and severity. Risk analyses are conducted during different operational phases and the format and time horizon for a risk analysis varies depending on the need for decision support.

The following sections first discuss the main types of risk analysis. We then discuss the Gullfaks C incident and the Deepwater Horizon disaster and highlight aspects of these events relevant to risk analysis. In both cases the event is described, then examples of the risk analyses that were actually conducted are investigated and finally the potential for improvement is examined.

The analyses of the two events are quite different. In the Gullfaks C incident it was possible to access the operative risk analyses. Consequently, we have a good understanding of how the risk analyses were conducted and the event can be discussed in detail. However, no risk analyses were available for the Deepwater Horizon disaster and the discussion is therefore based on an examination of the investigation report. This makes it difficult to be very clear about the use of risk analyses in the lead-up to the disaster.

Risk Analysis in Drilling and Well Operations in the Norwegian Offshore Industry

The following sections provide a brief description of risk analysis as it is currently used in the Norwegian offshore industry. Risk analysis is divided into three categories: strategic, qualitative design and operative.

Strategic risk analyses

Strategic risk analyses are primarily aimed at developing a safe design and safe operating procedures. Typically, the objective is to assess a proposed design or operation, to evaluate whether the risk level is acceptable and to identify potential risk-reducing measures. Strategic risk analysis takes a global perspective; it analyses the effect of a design or operation on the risk level for the entire installation. The analysis is usually quantitative, although it can also be qualitative. Strategic risk analyses tend to focus on technical aspects; operational input is usually limited to measurements of activity levels such as the number of offshore supply vessels visiting the installation, the number of crane lifts, the number of wells drilled and so on.

The part of the strategic analysis related to well operations usually assumes a blowout scenario from which, for example, event tree, fire and explosion models are developed in order to derive the risk picture. Blowout frequencies are usually based on information from the SINTEF Offshore Blowout Database (SINTEF, 2011). Strategic analyses do not normally focus on operational aspects of production such as the ability to maintain the integrity of two independent safety barriers or critical situation response times.

One example of a strategic risk analysis is the Total Risk Analysis that is performed for Norwegian offshore installations in order to determine the overall level of risk to personnel and to improve safety aspects of the design. Other examples are studies that are carried out before large-scale modifications. These studies are often based on the Total Risk Analysis, but change the inputs in order to obtain an updated risk picture. Typically, the responsibility for performing these studies lies with the design teams or with onshore staff responsible for technical safety on the installation. Strategic risk analyses are often related to major accidents, although in these cases the focus is more on individual systems than on the global risk level.

Qualitative design analyses

Qualitative design analyses are more detailed and more specific than strategic risk analyses and they typically focus on the system. The best-known examples are the HAZard and OPerability study (HAZOP) and the Failure Mode and Effects Analysis (FMEA). This type of analysis typically focuses on specific aspects of the system such as the blowout preventer or the mud system and the aim is to verify the design in detail in order to ensure safe and reliable operations. Like the strategic risk analysis, this type of analysis usually has a clear technical focus and is typically the responsibility of the design team or onshore personnel responsible for technical safety.

There are two types of qualitative design analysis. The first is a specific analysis conducted on a particular installation in order to establish the risk picture for that installation. The other is generic. An example is a safety barrier diagram analysis for generic well operations where the objective is to highlight risk factors for the type of operation. In this case, the end result is general procedures and descriptions of work for such operations.

Operative risk analyses

Operative risk analyses are different from strategic studies in almost every respect:

- They are typically qualitative studies that sometimes use a risk matrix to classify identified hazards or events and determine their acceptability.
- They are performed on a much more limited problem area. Typically this is an operation that is being planned or is about to be performed, or as support for a specific decision. This type of analysis may address major accident risk (although not necessarily), but the link to the installation's global risk picture (as described in the strategic analysis) is usually weak.
- Various groups may be responsible for these studies, including onshore planning/operations teams or offshore personnel who are responsible for carrying out the work.

This latter category typically consists of analyses carried out during drilling operations that are used to support decision-making.

The Gullfaks C Incident

This description and discussion of the Gullfaks C incident is based on information from two main sources: (1) Statoil's internal investigation report (Statoil, 2010) and (2) an audit of Statoil's planning for well 34/10-C-06A A by the Norwegian Petroleum Safety Authority (PSA, 2010).

Brief description of the event

The event is described in the operator's internal investigation report (Statoil, 2010: 6), which is quoted here:

> Well 34/10-C-06 AT5 on Gullfaks C was drilled in Managed Pressure Drilling (MPD) mode to a total depth of 4,800 metres. During the final circulation and hole cleaning of the reservoir section a hole occurred in the 13⅜" casing, with subsequent loss of drilling fluid (mud) to the formation. The casing was a common well barrier element,[1] and thus the hole in the casing implied loss of both well barriers. Loss of back pressure lead to influx from the exposed reservoirs into the well, until solids or cuttings packed off the well by the 9⅝" liner shoe. The pack-off limited further influx of hydrocarbons into the well. Both the crew on the platform and the onshore organisation struggled to understand and handle the complex situation during the first twenty-four hours. The well control operation continued for almost two months before the well barriers were reinstated.

Statoil's internal investigation team identified three direct causes of the incident (Statoil, 2010). The first immediate cause of the event was insufficient technical integrity of the 13⅜" casing resulting in a hole in the casing and thereby loss of a common well barrier element. A second cause was inadequate monitoring of the pressure in the C annulus between the 13⅜" casing and the 20" casing that allowed a leak in the 13⅜" casing to develop into a hole. Although the pressure had increased in the weeks before the incident, the increase was not noticed. A third contributing

---

1   The Norwegian Activity Instruction § 76 and the Governing Instruction § 2 require that there are two independent barriers protecting the well stream and that their status must always be known (author's comment, and not a direct citation from the internal investigation report).

factor was that the Managed Pressure Drilling (MPD) operation was begun and carried out without sufficient margin between the pore and fracture pressure, which made it more difficult to restore the barriers. Although this was not a direct cause of the event, it did influence how the event was handled.

The causes underlying the event are indicated by the internal investigation team's report (Statoil, 2010: 7):

> (i) The risk assessments performed in the planning phase were insufficient, (ii) insufficient risk evaluation during the execution of the MPD operation, (iii) insufficient transfer of experiences related to pressure control from the MPD operation in well C-01 in 2009, (iv) insufficient planning of the operation, (v) knowledge of and compliance with requirements, and finally (vi) MPD-knowledge and involvement of the Company's technical expertise.

This lack of, and deficiencies in, risk analysis and evaluation is a common finding in accident and incident investigations. A key question to be discussed later in this chapter is therefore how to improve risk analyses?

Use of risk analysis in well planning and execution

The investigation reports that followed the Gullfaks C incident make no explicit reference to either strategic risk or qualitative design analyses. Operative risk analyses are available, but it seems they were not linked to either the strategic or qualitative design analyses. This section provides an example of an operative risk analysis. It is followed by a discussion of the decision support it may (or may not) provide.

The operative risk analysis consisted of a set of brainstorming meetings where hazards and threats were identified and documented in a traditional risk register. Table 6.1 shows a sample of an entry in a risk register. Such analyses are usually conducted during planning of the well operation and there may be hundreds of entries in the risk register. The register should be updated continuously and is intended to be used as an operative decision support tool. However, it is not very clear from the available investigation reports and background documentation how these analyses were actually used in the decision-making process. Nevertheless, a few comments can be made based on our general understanding of the use of such analyses.

Typically, each entry in the risk register is allocated one of three levels: green, yellow or red. The main purpose of the analysis is to identify hazards and threats, assess their contribution to risk and identify risk-reducing measures for the entries categorized as yellow or red. If no appropriate measure can be found that can bring the risk level to green, approval for the activity is required from a higher organizational level. This means that activities categorized as yellow and red in the risk register are not show-stoppers, but they are considered critical and require more careful consideration. The investigation of the incident showed that several activities marked as yellow and red in the risk register were not approved at a higher organizational level, indicating risk management deficiencies in the planning of the well operation.

Table 6.1   Example of an entry in a risk register

Category	Assessment
Activity	(description of activity)
Consequence/loss	(description of consequence/loss)
Scenario description	(description of incident scenario)
Probability	(scale 1–5)
HSE impact	(scale 1–5)
Risk category	(green–yellow–red)
Risk improvement action	(description of risk treatment measures)
Status	(status of risk reduction action)
Responsible	(who is responsible for risk reduction action)
Due date	(deadline for implementing risk reduction action)
Adjusted probability	(scale 1–5, updated after risk reduction action)
Adjusted risk	(green–yellow–red, updated after risk reduction action)

Three principal questions arise regarding the operative risk analysis process:

1. To what extent is brainstorming able to detect all the threats and hazards relevant to an actual operation?
2. Does the format of an entry in the risk register make it possible to revise the assessment if conditions change (for example, the drilling method changes)?
3. How realistic is the assessment of the 'magnitude' of the threat or hazard?

A review of the risk register established during the operative risk analysis shows that a potential technical integrity problem concerning the 13⅜" casing was not identified and the answer to the first question is therefore not very promising. It is clear that when the event actually occurred the situation was made more complex as, without this clue, the crew handling the situation could not draw upon a suitable cause and effect model.

Grøtan (2010) distinguishes between two types of complexity: manifest and latent. Manifest complexity resembles complexity as described by Perrow – causes and effects are hard to understand but the outcome is not particularly surprising (Perrow, 1984). Latent complexity is distinguished by surprise (see Chapter 4 by Grøtan). The Statoil investigation into the Gullfaks C incident (Statoil, 2010) suggests that the event was characterized by latent complexity. The report shows that the crew were struggling to make sense of the situation and it was only when the increase in pressure in the C annulus became evident that any kind of understanding was reached. There also appears to have been failings in the risk analysis framework. The investigation report (ibid.) clearly states that the risk analysis process was far below acceptable standards.

Ways to improve

Here, we highlight the various constituent elements of a thorough risk analysis. If any of these elements had been applied, the potential casing failure would most likely have been identified.

The first, and simplest tool is to maintain a hazard log (or risk register) throughout the entire lifecycle of the installation. The purpose of the hazard log is to continuously document critical factors and conditions and actively use it in the risk management process. Integrity problems in the casing had been identified in November of the previous year. It would have been natural to use the hazard log to document this failure; consequently it would also have been almost impossible not to take this information into account in the operative risk analysis. A second way to identify problems with the casing would have been to use the qualitative barrier analysis (part of the strategic analysis) as input to the operative analysis, although this would have required a more detailed analysis than is usually the case when the blowout is used as the initiating event. If the strategic analysis was not detailed enough, a third way to reveal the casing problems would have been during the qualitative design analysis. The FMEA includes a systematic review of all technical failure modes. Casing failure is an obvious example of a failure mode.

Given the diverse methods that were available to reveal the root cause of the event, the real problem seems to be that the risk analyses were not conducted according to best practice, rather than that the issues were too complex for analysis within the framework.

The second critical question concerns the format of entries in the risk register and the extent to which they can be revised if conditions change. In the Gullfaks C incident well conditions meant that the drilling method was changed (from conventional to MPD) late in the planning process. The investigation report makes it clear that the Management of Change (MoC) process was inadequate (PSA, 2010, Statoil, 2010). However, the format of the risk register shown in Table 6.1 highlights that MoC is a particular challenge as the format does not provide for factors or conditions that change explicitly. In principle, *any* change requires an update of the *full* register – an impossible task. Therefore, in order to improve MoC, structural tools such as the risk register must be modified and greater emphasis should be given to factors that are likely to change (for example by adding a dedicated 'assumptions' column to the register). At the same time, the principles of Resilience Engineering suggest that it is important to strengthen the capacity of the organization to anticipate and adapt (see for example Hollnagel et al., 2011).

The third question that arises is how realistic is the assessment of the 'magnitude' of the threat or hazard? A risk matrix typically evaluates magnitude in terms of a probability assessment of whether an event will occur and the consequences if the event does occur. Such evaluations are typically arrived at through simple statements and documented in the risk matrix without further detailed analysis. As a screening mechanism, this is usually all that is needed to differentiate between red, yellow and green categories of risk.

However, a more detailed analysis might be required to fully understand the situation. For example, an entry in the risk register (see Table 6.1) may concern the function of the well stream mud column barrier. The mud column is one of the two barriers that prevent a major accident and potentially the total loss of the installation (see for example PSA, 2005 for a discussion of the Snorre A blowout). In the light of events at Snorre A an adjusted 'yellow'-category risk and the direct assignment of probabilities and consequences seems rather unrealistic for elements that may lead to a major accident. We therefore argue that, at a minimum, reference should be made to a more comprehensive risk analysis, that is, the qualitative design and/or the strategic risk analysis. It appears from the investigation reports that such links did not exist (PSA, 2010, Statoil, 2010). One reason for the missing link is that knowledge from the strategic risk analysis is codified into procedures. Such procedures are 'what to do' statements. They describe how to cope with, for example, common barrier elements. The problem with this codification of knowledge into procedures is that it loses sight of the reasons for the rules.

The Deepwater Horizon Disaster

This description and discussion of the Deepwater Horizon disaster is based on two principal sources, the National Commission on the Deepwater Horizon Oil Spill and Offshore Drilling Report (National Commission, 2011) and the Chief Counsel's Report (2011).

Brief description of the event

On the 20 April, 2010 the drilling rig Deepwater Horizon experienced a blowout followed by an explosion and fire. Eleven people were killed. The drilling rig sank two days later while the blowout continued for a total of 87 days.

The well had caused problems from the very beginning, with several kicks (an undetected influx of hydrocarbons) and lost circulation (loss of drilling fluids into the formation) incidents. The well was originally drilled by another rig (Marianas) but this had to leave the location following hurricane damage. Deepwater Horizon was then brought in to complete the drilling.

When the well reached approximately 5,545 metres below the seabed, mud circulation in the well was lost. This was the eighth time that this problem had arisen. However, as the well depth increased the problem was becoming worse and it was decided to stop drilling at just less than 5,600 metres, rather than the originally planned depth of around 6,160 metres. To prepare the well for production, it was lined with a 'long string' casing that ran from the seabed to the reservoir. Although this simplified well construction, it made cementing (which reinforces the casing and prevents the flow of hydrocarbons) more difficult. After cementing, the plan was to plug the well and temporarily abandon it.

During the preparations there was concern that the formation might continue to fracture and that cement would be lost (due to the earlier lost mud circulation incidents). Several decisions were therefore taken in order to reduce the pressure on the formation and consequently the probability of lost circulation. Cementing was completed and pressure tests were performed to confirm the integrity of the well. A negative pressure test (which simulates what will happen after the well is abandoned and determines whether the cement barriers will hold) verified that the cementing was successful and that there had been no influx of oil or gas from the reservoir into the well. However, this test had to be repeated several times and the procedure changed before it was eventually declared successful.

The crew then continued their preparations by replacing the heavy mud in the well with lighter seawater. During this process, there were several indications that the well was not stable and that oil and gas were flowing into it. However, these were not detected by the crew until mud, oil and gas started flowing out of the well and onto the deck. The crew attempted to divert the flow and shut the blowout preventer (thereby isolating the oil reservoir from the surface) but this failed and shortly thereafter an explosion occurred.

The Chief Counsel's Report (2011: x) states that, 'The root technical cause of the blowout is now clear: The cement that BP and Halliburton pumped to the bottom of the well did not seal off hydrocarbons in the formation.' The exact reason for this may never be known, but several factors that increased risk were identified:

- lost circulation forced engineers to plan a 'finesse' cement job;
- the cement slurry was poorly designed and unstable;
- the abandonment procedure called for a severe underbalancing of the well;
- the result of the negative pressure test was misinterpreted;
- signs that hydrocarbons were entering the well were missed;
- the blowout preventer failed to isolate the hydrocarbon flow.

It was concluded that most of the technical failures could be traced back to management errors by the companies involved in the incident. Furthermore, it was concluded that the regulatory structure in place did not adequately address the risks of deep-water drilling projects. Many critical aspects of drilling operations were left to the industry to decide, without regulatory review. Examples of this include the fact that there were no regulatory requirements related to negative pressure testing or testing of cement stability.

The following sections specifically address the use of risk analysis. They discuss whether risk analysis is a suitable tool for identifying hazards and accident scenarios, assessing the magnitude of the risk and supporting decisions which may increase or decrease risk. We look at both how risk analysis was used in the period preceding the accident, and also how it could have been used.

Use of risk analysis in well planning and execution

The use of risk analysis is discussed in the Chief Counsel's Report (2011) and we will highlight some of its findings. However, the report does not go into detail about how the risk analyses were performed and the information is therefore to some extent supplemented by general knowledge of common practice in the industry.

Well planning included a rigorous peer review process and a risk assessment. This can be seen as part of the strategic risk analysis. The report does not describe whether this process was performed in accordance with the requirements, but we will assume that this was the case. It therefore follows that risk had been evaluated and that a reasonable balance between risk and cost/operation had been found.

More relevant to the accident is the fact that after well operations had started, operative risk analyses were only used to a very limited extent. In this phase, the requirements for formal risk assessment were tied to the MoC process. Whenever this was initiated, a risk analysis was required. However, it was only initiated when there were changes to the well plan, not well procedures. The exact requirements are not clear, but in practice it meant that the MoC process was initiated only in three cases:

- when it was decided to change from a 16″ to a 13⅝″ casing string;
- when the total depth of the well was reduced;
- when it was decided to employ a long string rather than a liner.

The Chief Counsel's Report (2011: 43) concludes that, 'Almost every decision […] identified as having potentially contributed to the blowout occurred during the execution phase.' This means that a large number of decisions were not analysed systematically.

The risk analyses that were performed in the execution phase are not described, but it is likely that they would be 'risk matrix'-type analyses where the probability and consequence of failures/events are classified in broad categories and the risk level is determined based on the combination of the two. Although the exact format is not known, it is likely to be similar to the format used at Gullfaks C.

In short, only a few decisions (changes) made during the execution phase were actually analysed and there were clearly shortcomings in these analyses. Another significant shortcoming is that the analyses underestimated the importance of the various types of hazards. The well had experienced both kicks (precursors to a blowout) and lost circulation incidents, but in the subsequent decisions greater weight was given to avoiding lost circulation than avoiding a kick or blowout. In the end, lost circulation was avoided, but a blowout was not. Of course we will never know whether lost circulation would have occurred and the blowout would have been avoided if different decisions had been made, but lost circulation clearly has less severe consequences than a blowout.

Another important issue is the fact that the risk analyses were carried out in isolation and decisions based on these analyses were taken in isolation from other decisions that affected the same risk. This is a weakness with operative risk analyses, which typically focus on small parts of the system, a single operation or a single decision and only assess the risk associated with it. Each individual decision only looks at how the risk level will change depending on the available alternatives. Often, this is a comparison between highly likely benefits (saving cost and or time) and uncertain and highly unlikely negative effects (blowout).

Although a strategic risk analysis provides an overview, it is not sufficiently detailed to reflect changes in risk level that are the result of many smaller changes. There is therefore no way to track how the risk level increases or decreases over time. In a situation where many changes are made (as was the case for the Deepwater Horizon well) this can lead to an increasing probability of an event occurring. This increase may not be recognized until it is too late.

Situations like this typically use a risk matrix to assess risk. However, in most cases small-scale decisions do not change the risk matrix or the overall risk picture. This is because probabilities and consequences are defined very broadly. Typically, probabilities are defined in orders of magnitude. This means that in order for a hazard, an event or a failure to move from one category to the next, there must be a ten-fold increase in probability. Even a two or three-fold increase will not necessarily be reflected in the matrix. Moreover, it is difficult to detect changes in risk level because there is no easy way of showing that the risk is approaching the next 'box' in the risk matrix, assuming it is updated throughout the decision chain. A fuller discussion of some of the weaknesses of risk matrices is given by Cox (2008).

These processes are illustrated in Figure 6.1. The solid arrows illustrate when each decision is made using the risk level calculated from the qualitative risk assessment as a baseline.

The figure shows that some decisions increase risk and others decrease risk, but the starting point for each decision is always the same. The cumulative effect of many decisions is shown by the dotted lines and arrows.

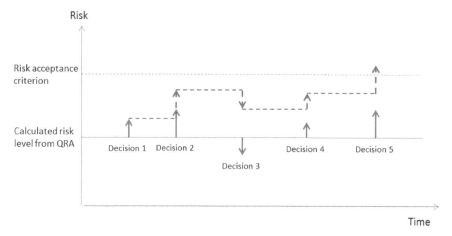

Figure 6.1   Effect of cumulative decision-making

Another factor that must be taken into account that adds to the complexity of decision-making in operational situations is the fact that most decisions will only have a marginal effect on the risk of a major accident. There are many possible causes of accidents and complex causal combinations, which means that a multitude of factors can influence the probability of an incident. The defence-in-depth principle means that changes to almost any individual factor will only have a small impact on the risk level. The result is that even relatively major decisions, such as the removal of a barrier system, will not necessarily cause a large increase in risk. This makes it even more difficult to resist the temptation to exchange a large and certain benefit for a small increase in a very unlikely consequence.

Decision support 'failure modes'

The discussion in the previous section suggests that it may be useful to try to outline the ways in which risk analysis can fail as a decision support tool. There seems to be at least three ways in which this can happen:

1. The analysis may fail to identify a scenario that actually occurs. For example, the risk analysis does not include the possibility of a blowout or the potential failure of the cement barrier. Although the investigation reports from the Deepwater Horizon incident do not provide details of the risk analyses that were performed it is highly unlikely that these risks were not assessed, particularly as blowout is the single most important contributor to risk in drilling operations. This point relates to the discussion of manifest versus latent complexity in Chapter 4.

2. The risk analysis may identify the possibility that an accident may occur, but fail to both realistically assess the risk level and provide a solid basis for decision-making. It is likely that this was the case in several of the decisions that were made in the Deepwater Horizon disaster. The team appears to have focused on the issue of lost circulation and solving this problem was given priority over avoiding an influx of hydrocarbons into the well.

3. The risk analysis may not be performed at all, either because no risk is foreseen or because it is not considered necessary. The previous discussion has highlighted that in the events leading up to the disaster, most decisions were not subject to a formal risk analysis. Although risk analysis is likely to have been an informal part of decision-making, it is impossible to be sure from the evidence presented in the investigation reports.

In the Deepwater Horizon disaster, it was primarily the second and third types of failures that occurred. However, the fact that a risk analysis may not have been carried out does not really contribute to a discussion of the suitability of risk analysis as a tool for decision support. We will therefore focus on the first and second types of failures. An examination of the general categories of accidents shows that it is rare for a systematic, strategic risk analysis to miss a scenario completely. In the qualitative risk assessments carried out for Norwegian offshore installations, a blowout (including a blowout during well completion) is one of the standard event categories. However, strategic risk analyses do not include detailed descriptions of specific scenarios and all potential causal paths. One reason for this may be that historically the purpose of this analysis was to verify, in the design phase of the project, that the level of risk was acceptable. They therefore focused on technical rather than operational issues. Nevertheless, the methods that were developed to carry out these analyses are still being used. The second type of failure – misjudging the risk level – also appears to have been a factor in the Deepwater Horizon disaster, in particular in the operative risk analyses.

Results and Ways Forward

This section summarizes the main results and gives some general recommendations regarding the use of risk analyses.

Failure to identify potential scenarios

A frequent criticism of risk assessment is that the risk analysis fails to identify a situation that actually happened. It is important to realize that the purpose of a risk analysis is not to identify and explicitly address every possible chain of events that may lead to a severe accident. Instead, it is to provide decision support for the most important issues. Therefore we argue that greater focus should be placed on the factors that influence risk rather than examining each and every causation path.

A failure to identify potential scenarios can occur at both the strategic and operational level. Although the aim of the strategic analysis is to identify potential scenarios, they are representative and not necessarily examined in great detail. Moreover, it is obviously impossible to analyse all possible accident sequences. In terms of qualitative design and operative analyses FMEA and HAZOP would appear to be able to identify key issues. The discussion of the Gullfaks C event suggests that more structured use of these techniques would have revealed many of the problems that proved to be the main causes of the incident. Obviously, the question remains of whether the tools used in the identification process need to be improved, but our impression is that the existing tools are appropriate.

Weak links between levels of analyses

We argue that there is a weak link between the strategic and operational levels of risk analysis. This gap between the two levels means that important information does not flow between them and the basis for decision support is weakened. Strategic risk analyses do not examine causality in detail and do not address the question of the safety barriers to be put in place to prevent accidents occurring. As their primary focus is how to prevent accidents, they fail to provide a sufficient foundation for the subsequent operative analyses. We therefore recommend the development of comprehensive scenario models that can be used as a common platform for all three types of risk analyses.

Chapter 13 is an illustration of scenario modelling. In this approach the model's parameters (for example, failure frequencies and human error probabilities) are assessed. Parameters are affected by various Risk Influencing Factors (RIFs). For example, the failure frequency parameter may be affected by RIFs such as the age of the well, the quality of inspection equipment and structural integrity management. Resilience Engineering emphasizes anticipation as one of the abilities of a resilient organization (Hollnagel et al., 2011). Anticipation is another example of a RIF and in the Gullfaks C event, it seems to have been lacking. There was no systematic check of the pressure in the C annulus and consequently a small leak was able to develop, unnoticed, into a hole.

This approach would be more useful if it was possible to be specific about what anticipation related to. For example, a distinction can be made between the anticipation abilities of the organization in the planning phase and during operations. In the planning phase anticipation relates to the preparation of the risk register, while in the operative phase it includes the impact of a loss of casing integrity. In the latter case, information about well conditions would then provide various cues to be monitored, for example, pressure build up or mud loss. Anticipation of critical situations could be related to time pressure or the ability of distributed teams to create mental models of the situation. Other, less direct RIFs include common shared awareness (CSA), the local knowledge of operational personnel and access to external sources of knowledge (for example drilling support centres).

A more extensive discussion of risk influence modelling can be found in Vinnem et al. (2012).

Failure to realistically assess the risk level

At least three issues can be identified in relation to the failure to make a realistic risk assessment:

1. Decisions are taken in isolation and do not take into account relevant earlier decisions.
2. 'Failing' by a decade in the operative risk assessment.
3. Inability to assess the effect of changes.

These issues can become apparent in the use of the risk matrix and when a small increase in a very low probability of a very negative outcome is compared with a highly likely benefit.

Another benefit of the proposed integrated model is that it provides a more realistic risk picture, which is based on a shared understanding of a scenario. This can be used as input to decisions about risk tolerability and improvement measures. For example, the model makes it possible to visualize potential risk reductions if the anticipation abilities of the organization were strengthened. The model would also be valuable in qualitative design and operative analyses as it would make it possible to visualize potential accident scenarios and link the findings of the operative analyses to the overall risk picture.

A set of scenarios, focused on common barrier elements would be valuable in assessing the risk level documented in the risk register. For example, in the Gullfaks C incident it was apparent that the assessment of the magnitude of the risk was questionable. If each entry in the risk register was linked to a scenario it would be easier to visualize the total impact of an individual element. This would also enable the various entries in the risk register to be linked to each other and not simply presented separately.

Would Integrated Operations make a difference?

It is beyond the scope of the current discussion to give a comprehensive assessment of what would have happened had

Integrated Operations (IO) been implemented in the two incidents presented in this chapter; nevertheless some observations can be made. One of the main objectives of IO is to improve complexity management. One way to achieve this is to use onshore support centres that are responsible for coordinating the three types of risk analyses. Such support centres would be distant from actual operations and maintain a high-level overview of ongoing activities. With appropriate training, the personnel in such centres should be able to create mental models of risk based on an integrated set of risk analyses created on the fly in a critical situation. Critical events are characterized by a high level of complexity caused largely by a lack of understanding. In these circumstances, it is reasonable to assume that proper implementation of IO would improve the situation given the access it provides to up-to-date risk analyses. At the same time, there is no doubt that better understanding is achieved by a thorough risk analysis in the planning phase and active use of the risk register (hazard log) during operations. Whether this actually means that the right people actively access and use the risk register and analyses in decision-making is another issue.

The investigation of the Gullfaks C incident highlighted the fact that the organization's existing drilling expertise was not fully used in the planning phase. IO offers new ways of working and a key aspect of the approach is the more coordinated, systematic use of expertise in the various critical phases of operations (see Chapter 13 for further discussion of these issues).

Conclusions

This chapter has discussed the practice of risk analysis. Risk analysis can be grouped into three types: strategic, qualitative design and operational. Each type differs in terms of granularity, the techniques deployed and the objective of the analysis. Whatever the type, risk analysis can fail to support decision-making in several ways: the scenario was not identified, the scenario was identified but the risk level was wrongly assessed or the risk analysis was not performed at all. This chapter used these typologies to analyse two accidents from the oil and gas industry and examined the risk analyses carried out, the possible failures and their contribution to the accident.

The analysis focused on the Gullfaks C and Deepwater Horizon incidents. One important finding is that current practice in the industry is affected by many threats and hazards which could be mitigated by measures identified in a risk analysis. For example, in the Gullfaks C event the root cause of the incident was identified, but not addressed in the operative risk analysis. We argue that better integration of the various types of analyses and the systematic use of a hazard log during the entire life of an installation would provide better support for hazard identification during operations.

Our conclusions are based on examples from drilling and well operations. These operations are highly dynamic and the status of a well can change dramatically in a very short period of time. On the one hand this makes these situations somewhat unusual compared to other offshore operations. On the other hand, the types of risk analyses that are applied and the associated failure modes are similar for most types of high-risk operations, even if the various elements are given a different weight. It is therefore likely that the conclusions reached here can be applied to other critical operations, such as maintenance and modification operations on process plant equipment.

References

Chief Counsel's Report. 2011. *Macondo – The Gulf Oil Disaster*. National Commission on the BP Deepwater Horizon Oil Spill and Offshore Drilling. Available from http://tinyurl.com/3eyru37.

Cox, L.A. 2008. What's Wrong with Risk Matrices? *Risk Analysis*, 28(2), 497–512.

Grøtan, T.O. 2010. *Risk Governance in Complex Environments – Making Sense with Cynefin*. Paper presented at Working on Safety: 5th International Conference. Røros, Norway, 7-10 September 2010.

Hollnagel, E., Jean Pariès, J., Woods, D.D. and Wreathall, J. 2011. *Resilience Engineering in Practice*. Farnham: Ashgate.

National Commission. 2011. *Deep Water – The Gulf Oil Disaster and the Future of Offshore Drilling*, Report to the President from the National Commission on the BP Deepwater Horizon Oil Spill and Offshore Drilling, January 2011.

Perrow, C. 1984. *Normal Accidents: Living with High-risk Technologies*. New York: Basic Books.

PSA (Petroleum Safety Authority). 2005. *Investigation of Gas Blowout on Snorre A, Well 34/7-P31A, 28 November 2004*. Petroleum Safety Authority Norway.

PSA (Petroleum Safety Authority). 2010. *Audit of Statoil's Planning for Well 34/10-C-06A A. Petroleum Safety Authority Norway*. Petroleum Safety Authority. Available at http://tinyurl.com/746h79f.

SINTEF. 2011. *SINTEF Offshore Blowout Database*. Available at http://www.sintef.no/home/Technology-and-Society/Safety-Research/Projects/SINTEF-Offshore-Blowout-Database/.

Statoil. 2010. *Brønnhendelse på Gullfaks C* [Internal investigation report by Statoil after a well event on Gullfaks]. Available from http://www.statoil.no/.

Vinnem, J.E., Bye, R., Gran, B.A., Kongsvik, T., Nyheim, O.M, Okstad, E.H., Seljelid, J. and Vatn, J. 2012. Risk modelling of maintenance work on major process equipment on offshore petroleum installations. *Journal of Loss Prevention in the Process Industries*, 25(2), March 2012, 274–292.

Chapter 7

Assessing the Performance of Human–machine Interaction in eDrilling Operations

Denis Besnard

It has been clear since the 1980s that the introduction of automated systems for control and supervision tasks has significant drawbacks resulting from human–machine interaction (HMI). However, there have been few attempts to create a method to assess either the existence or the effect of these drawbacks. This chapter describes such a method and documents the various engineering steps. This chapter first surveys the literature on the drawbacks of automation and then describes how these can be converted into a set of HMI assessment criteria. These criteria are then engineered into a prototype method and finally deployed in a case study in an eDrilling scenario. The results show that the prototype method can be used to assess the performance of HMI in control and supervision tasks.

Automation and Complex Systems

The automation of complex systems does not mean that human intervention is no longer needed in control and supervision tasks. In fact the opposite is true; the fact that systems can enter exceptional and unforeseen states makes the role of humans all the more important. Given that the role played by automated systems has become increasingly business-critical it follows that, more than ever before, humans are needed. This is one of the ironies of automation identified by Bainbridge (1987). However, the essential contribution that humans make to the control and supervision of complex and dynamic tasks can be jeopardized by the system itself. This chapter will investigate this paradox.

Control and supervision refers to the mental activity that humans use to manage dynamic processes – from flying an aircraft to controlling production chains. Put simply, it is an activity where environmental cues are used to assess the current state of a given process and decide whether something has to be done or not, and if so when and how, and so on. To take the example of piloting an aircraft, pilots have a flight plan that they try to follow. The plan consists of various way points, altitudes and speeds that govern the progress of the aircraft. A simple view is that the role of the pilot is to execute the actions that make the aircraft follow the flight plan and reach its destination. For the most part, these actions now involve keying data into the on-board flight management computer.

The computerization of the HMI and the high level of automation that came with it were initially thought to guarantee reliability. However, in the mid-late 1980s (when information technology (IT) was becoming more and more prevalent in the world of process control) some important issues started to surface. Concomitant with the technological shift, new forms of accidents began to appear in which pilots lost control of their aircraft under normal flying conditions. Such events began a reconsideration of the role of automation in control and supervision tasks (Besnard and Baxter, 2006). Research was initially pioneered by the safety-critical commercial aviation industry, which was badly affected by such events. Today, HMI automation is a paramount concern in many domains and extends well beyond the field of aviation, for example to the control rooms of oil and gas pipeline systems (Meshkati, 2006) or collaborative decision-making (Pierce and Salas, 2003).

This chapter addresses the issue of HMI automation in the specific context of offshore oil and gas drilling operations and focuses on a tool known as eDrilling. eDrilling is one example of an integrated drilling simulator system that has been developed and tested in some oil and gas companies. Such tools simulate drilling operations by offering a 3D visualization and provide control from a remote drilling expert centre supported by real-time data. It is used in a case study to develop a prototype HMI assessment method. This method aims to identify factors that might develop into a loss of situation awareness on the part of operators, particularly where there might be insufficient understanding of automated behaviour (Endsley, 1996). The eDrilling tool is safety-critical, extensively computerized and includes complex control interfaces. These features imply that operators using it can expect to face the same challenges that have created mishaps in other industries and make the assessment of eDrilling worthwhile. Hollnagel and Woods (1999: 346) express the idea in another way: the purpose of the evaluation is to assess the 'match between the system's image and user characteristics on a mental or cognitive level'.

The following section reviews some of the drawbacks of automation. Next, these are converted into a set of interface-centred evaluation criteria for eDrilling. Finally, this material is integrated into a prototype method which aims to assess HMI performance in automated environments such as Integrated Operations (IO) (see Chapter 2 for a description of IO).

Six Drawbacks of Automation

The literature provides several classifications of automation levels, dating back to the seminal study of Sheridan and Verplank (1978). This study influenced later work such as Parasuraman et al. (2000), which lists ten levels of automation. These ten levels include situations where the operator:

- has full and direct control over the task;
- has full, but indirect control over the task through an automated interface, decides which actions to perform and receives feedback; and
- interacts with the task indirectly through an automated interface, does not always decide which actions to perform and does not always receive feedback.

These various levels of automation are summarized in Figure 7.1.

Such a classification makes it possible to explicitly specify the allocation of functions between humans and machines. It has a significant bearing on the content of the supervision and control task and partly determines what the human operator will have access to, in order to build their mental representation

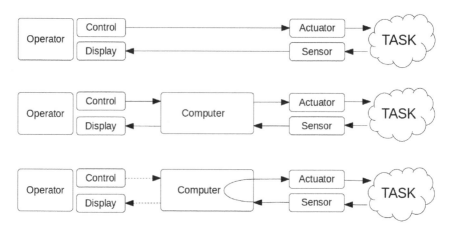

**Figure 7.1   Representation of three levels of automation of a
                computerized control task**

Source: Adapted from Parasuraman, Sheridan and Wickens, 2000

of the process. Moreover, the classification (despite its age) is a
reflection of the drawbacks of automation, as it is generally the
case that the greater the extent of automation, the more difficult
it is for an operator to detect a loss of situation awareness. This
makes the effects on the control and supervision task all the
more severe. In view of the consequences of a loss of situation
awareness, it makes sense to address the issue of the drawbacks
of automation in some detail. This discussion may eventually
help the operator find an answer to the key question of control
and supervision, namely, 'Why is the system doing what it does?'
(Boy, 2005), which is the focus of the following subsections.

The following sections present and discuss six HMI-related
drawbacks resulting from the automation of systems. They draw
heavily upon the work of Abbott et al. (1996) who provide an
extensive discussion of automation issues, particularly aboard
modern aircraft. Although the six examples given here do not
originate from the oil and gas domain, it is worth noting that the
two domains have many common features. The eDrilling operator is
just as remote from the process as the pilot of a fly-by-wire aircraft.
In both situations the control cues have changed, the system has
become more complex and autonomous, and anticipation is a key
issue. With these similarities in mind, it is assumed (in the absence

of a full demonstration) that the difficulties found in eDrilling are comparable to those found in traditional HMI-critical domains.

Remoteness of control

Modern interfaces tend to separate operators from direct contact with the process they are controlling. Activity becomes symbolic: rather than direct physical information the process is controlled by a set of parameters. This can be seen in nuclear power station control operations, air traffic control (Figure 7.2) and train traffic control (Crawford et al., 2010) for example.

The fact that processes can only be indirectly manipulated means that performance control relies on the operator having an accurate mental representation of a symbolic set of information. This symbolic representation is not necessarily a source of

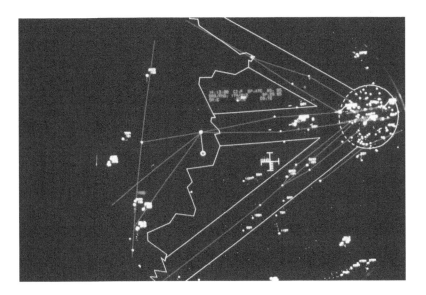

Figure 7.2   Air traffic control display of the aircraft corridors at West Berlin airport in 1989

Source: Wikimedia Commons

mishaps per se. However, it can become a challenge for operators when control and supervision tasks switch from being direct and analogue, to indirect and symbolic.

In the context of IO, remote operations are only marginally implemented. The current situation may be due to a lack of feedback from experience and insufficient technological maturity. As the technology is deployed more extensively, it is reasonable to expect that remote operations will become common practice.

Change of control cues

Automation does not simply mean that a machine performs part of a task. It also changes the nature of the work carried out by humans and what they use as control cues. In aviation, for example, the mechanical commands that controlled surfaces such as flaps were replaced by electronic commands. Consequently, pilots were no longer physically connected to these control surfaces by cables and pulleys and they no longer received the haptic feedback provided by vibrations. This was a significant change as pilots lost a direct cue about airflow over the wings of their aircraft. The shift to electronic commands proved to be such a handicap for pilots that simulated vibrations had to be built back into modern control columns.

The problem of changes in control cues that follow the introduction of remote or indirect control technology is not unique to aviation. It was also shown to be an issue for high-temperature furnace operators when steelworks factories implemented process control rooms. Operators lost direct cues related to the temperature, colour and texture of the molten metal. Later, the move from electro-mechanical to digital displays caused a similar change in the control rooms of nuclear power plants (Figure 7.3). Before the introduction of digital displays a change in a parameter would cause a dial to click, thereby providing auditory information that something was happening. With digital technology, these sounds disappeared, which meant that operators could only supervise changing control parameters by sight.

Figure 7.3   A set of electro-mechanical panels in a nuclear power plant control room

Source: © 2009 Yovko Lambrev. Creative Commons

Opacity

Automation (especially computerization) has made it difficult for the operator to track the controlled process. The situation is analogous to a black box. To take the example of Babbage's difference engine (Figure 7.4) it is clear that the various cogwheels, levers and drums work together to produce the result of a computation. However, once digitized into a pocket calculator (Figure 7.5) the process is no longer visible.

Without prior knowledge of the internal design of the calculator, only the input interface (the keyboard) and the output interface (the monitor) provide the user with information from which they may be able to infer how calculations are actually done. The same is true for any machine or system that relies on digital technology. This opens the door to automation 'surprises', a term which refers to decisions taken by machines

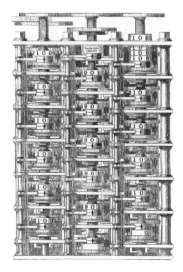 

Figure 7.4   Charles Babbage's difference engine

Source: Portion of Babbage's difference engine, *Harper's New Monthly Magazine* (December 1864), 30(175), 34.

Figure 7.5   A pocket calculator

Source: Wikimedia Commons

that are not consistent with the understanding that crews or operators have of the control task. Such an event took place in 1995 aboard a Boeing 757 aircraft approaching Cali airport (Colombia). The crew keyed the first letters of a navigation beacon into the flight management system (FMS). Unknown to the crew, the FMS wrongly self-completed the name of the beacon. Consequently, the aircraft turned towards the incorrect beacon, located on a trajectory that did not follow the original descent path. By the time the crew detected that they were on an incorrect heading and corrected their trajectory, they were in a mountainous area and the aircraft ploughed into a mountain. Everyone on-board perished in the crash (ACRC, 1996).

Complexity

As automation becomes more extensive, complexity increases (see Rauterberg, 1996 for a definition). This not only means that there are more functions, but also that the number of possible interrelations between functions increases. For example, the control unit of modern car engine receives a large amount of digital information (engine load, engine speed, composition of exhaust gases). This information is used to control the functioning of the engine (for example, by adjusting the composition of the fuel/air mixture).

These automated functions operate without intentional input from the operator (the driver). They also have cross-compensation capabilities, which mean that a manual change to one of them will not necessarily change the overall state of the system (that is, the functioning of the engine). Such features are efficient under nominal conditions. However, in the case of failure where there is a need for manual tweaking, or in an emergency situation, the complexity inherent in the system makes adjustment difficult. First, greater complexity makes it increasingly difficult to diagnose the cause of a failure as more functions create more possible failure combinations. Secondly, a manual override can be made impossible by design (for example, it can be impossible to repair an engine control unit).

Mode confusion and system autonomy

A mode is a pre-programmed, selectable set of instructions that configures the behaviour of the system it controls. Modes make it possible to streamline an interface by assigning more than one function to a given control and they are paramount in modern control and supervision systems. In aviation for example, the same device enables pilots to manage descent speeds in two different ways (Figure 7.6): a push button in the middle of the Flight Control Unit (FCU) selects the descent mode: vertical speed (V/S) or flight path angle (FPA). A rotator button is then used to input descent data in both modes. This arrangement has the advantage of triggering several actions at once, thereby facilitating interactions with the system. However, as Sarter and

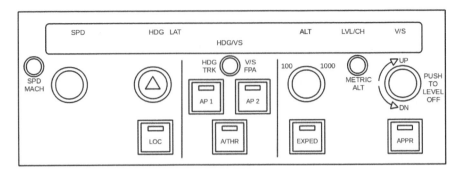

Figure 7.6   **A simplified representation of the control interface
to the Flight Control Unit (FCU) of an Airbus A320**

Woods (1995) have pointed out, system autonomy can make it
difficult to interact with modes as it can cause modes to change
indirectly. An example of this from the aviation industry is a
system-generated change in a flight parameter (as opposed to a
manual intervention by the pilot) that triggers a mode awareness
issue affecting the crew.

The crash of an A320 aircraft on Mont Sainte Odile in 1992
(METT, 1993) and the accident involving an A300 aeroplane in
Nagoya, Japan (AAIC, 1996) are both examples of incidents where
the wrong mode was selected. Consequently, in both cases the
pilots could not understand why the aircraft was not behaving
as expected. Both of these accidents resulted in many lives being
lost, although both aircraft behaved exactly as designed. This
paradox is known as 'controlled flight into terrain'. This term
highlights both a) the absence of technical failure as the cause
of the accident and b) the inconsistency between the operator's
mental model and the actual flight situation.

Anticipation

In dynamic control situations such as driving a car, anticipation
is required in order to foresee a range of possible future states,
and prepare control plans and strategies. In particular, the
speed of motion, the complexity of the control task and external
conditions determine the extent of the anticipation envelope. For
example, when driving at high speed the information used by the
driver as input to the control task has to be extracted far ahead

of the position of the vehicle, in both time and space. The same applies to high-tempo control and supervision tasks. This is true to such an extent that the proficiency of operators in controlling such tasks can be indirectly assessed by their ability to predict future system states.

This section has addressed a number of the drawbacks related to the automation of process control and supervision tasks and has demonstrated how HMI performance can degrade as a result of these issues. This list of drawbacks provides the input for the prototype HMI assessment method described later in this chapter. For now, the following section discusses the main features and operational environment of eDrilling.

Remote Control and Supervision in Integrated Operations: eDrilling

The toolbox for offshore drilling and well operations is likely, in the near future, to include integrated drilling simulators such as eDrilling. This tool is expected to be deployed within an IO context, which provides the necessary technical environment.

Integrated Operations

In Chapter 2, Eirik Albrechtsen provides an extensive description of IO. The interested reader should therefore refer to this chapter for further details. However, in the present chapter the defining properties of IO described by Albrechtsen and Besnard (2010) are used, summarized here as:

- the use of IT and digital infrastructure to enable new work practices;
- greater capture of offshore performance data;
- the use of real-time data to monitor operations across geographical and organizational borders;
- the use of collaborative technology to link various actors; and
- access to expert knowledge.

Among its many benefits, IO enables remote operators to control and supervise physical processes such as drilling. In this operational environment, human controllers make process-related decisions on the basis of symbolic cues, in dynamic and highly computerized situations, working in collaboration with remote colleagues. This setting forms the context for the implementation of eDrilling.

eDrilling

eDrilling is a software-based control and supervision system that enables physically remote teams to collaborate on drilling operations in oil and gas fields. eDrilling combines:

- a 3D dynamic drilling simulator;
- an automatic quality check and correction of drilling data;
- real-time supervision technology for the drilling process; and
- a diagnosis of the state of drilling and conditions (Rommetveit et al., 2008a).

An eDrilling control interface is shown in Figure 7.7; it can also be implemented on a flat screen and used as a shared display. At the present time there is only one implementation of eDrilling in offshore drilling operations in Norway, although other similar systems exist in the global petroleum industry. These systems are generally known as integrated drilling simulators.

A core feature of such systems, including eDrilling is the use of real-time data processing, which enables real-time communication between onshore and offshore drilling actors and the shared supervision of drilling processes. Real-time data processing also enables simulations to be carried out. One application of this is the ability to make a diagnosis of the drilling state and conditions (for example, the temperature profile and friction). Simulations also enable the generation of a) early warnings of upcoming unwanted conditions and events; b) test drilling plans; and c) drilling scenarios.

The ability to simulate drilling is an important HMI benefit as it provides operators with a model of potential future drilling conditions. In theory, this places them ahead of the process, which enables them to anticipate future states and proactively adjust

Figure 7.7   A computer-generated image of a potential eDrilling implementation

Source: © eDrilling Solutions

drilling decisions. This is a major asset in the control of processes that can be either difficult or impossible to pause or stop.

A full description of the eDrilling system is outside the scope of this chapter and further details can be found in Rommetveit et al. (2007) and Rommmetveit et al. (2008a, 2008b). However, it should be noted that although these references provide a fine-grained description of eDrilling, there is little discussion of the HMI challenges related to automation or, more generally, of the mishaps that may occur as a result of the introduction of automation into drilling operations. These issues are the subject of the next section, which presents a prototype method for HMI assessment. Given the focus of this chapter on HMI, the method aims to address the question of safety-critical HMI concerns related to eDrilling.

A Human–Machine Interaction Assessment Prototype Method

This section presents the prototype method, which consists of four steps:

1. select the system functions to assess;
2. develop assessment questions on the basis of the drawbacks of automation;
3. assign scores and produce a graphical representation; and
4. assign the set of scores to a risk class.

Select the system functions to assess

HMI assessment uses advanced models and methods to evaluate situations such as the effect of location on a given display (Xiaoming et al., 2005) or information searching performance (Ha and Seong, 2009). Here, we will only address a few of the many possible assessment dimensions.

Parasuraman et al. (2000) and Parasuraman and Wickens (2008) propose that the functions of automated systems can be divided into four classes, namely: information acquisition, information analysis, decision selection and action implementation. In each of these classes the extent of human–machine collaboration is distributed over a spectrum that ranges from low to high, as shown in Figure 7.8. The following eDrilling recommendations rely on the four function classes:

- *Information acquisition should be highly automated.* Sensors integrated into an IT system make it possible to quickly extract a wide range of data with a lower failure probability than humans are capable of.
- *Information analysis should be collaborative.* Automation should be used to narrow down the set of alternatives, but leave the final decision to humans.
- *Decision selection should usually be done by humans* although computers can assist humans in knowing the consequences of their actions (this is a current feature of eDrilling).
- *Implementation of actions can be almost entirely automated* with humans playing a supervisory role.

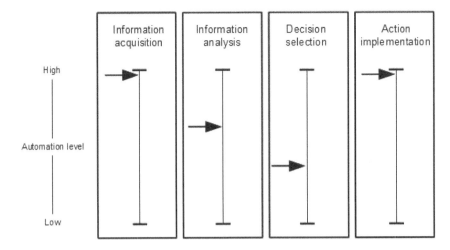

Figure 7.8   Recommended automation levels for eDrilling
Source: Adapted from Parasuraman et al., 2000

The work of Fitts (1951) demonstrated that machines and humans have different capabilities, and identified what these different capabilities are. Therefore, it is unsurprising that the degree of automation should vary across functions. The above classification shows that HMI is most critical in *Information analysis* and *Decision selection*. It therefore follows that an HMI assessment of these two classes of functions would be useful.

Develop assessment questions from automation drawbacks

This section describes how to convert the six drawbacks of automation described earlier into simple HMI-centred assessment questions.[1] It should be noted that producing an appropriate set of questions requires a detailed understanding of the system, its various functions and how these functions come into play in the performance of a drilling task. Typically, this requires the assistance of eDrilling specialists (designers and operators). However, given that this chapter only aims to present a prototype, this phase has been omitted.

The eDrilling scenario presented here describes a case where only one offshore operator pilots the eDrilling simulator and takes drilling-related decisions. Although this may not be the typical eDrilling scenario it nonetheless describes conditions that can be considered

---

1   This section draws heavily on the work of Abbott et al. (1996).

the most safety-critical: the process is controlled with no direct feedback (noise, vibration and so on) and there are no collaborators available to provide suggestions for recovery in the case of a mishap.

The six drawbacks of system automation and their related eDrilling assessment questions are shown in Table 7.1. The six generic drawbacks have been used to generate two distinct types of questions, namely:

- Questions assessing HMI performance: Can this situation occur? Is this problem present?
- Questions assessing safety: Can this drawback affect a safety-related task? To what extent does eDrilling provide a way to recover from this drawback?

Table 7.1   HMI drawbacks and corresponding questions

Drawback	eDrilling Assessment Questions
Remoteness of control	HMI performance
	Can the operator accurately mentally represent what is happening at the rig? Are there cases where the operator's mental model can differ from what is happening at the drilling site? Can the operator detect that their mental representation is not consistent with what is happening at the drilling site? Are there safety-critical consequences when the operator's mental representation is not consistent with the reality at the drilling site?
	Safety
	To what extent are safety-critical eDrilling tasks exempt from issues related to remoteness of control? To what extent can eDrilling help the operator readjust their mental representation?
Change of control cues	HMI performance
	Are there any discrepancies between analogue and eDrilling control cues? Can analogue cues be simulated by the eDrilling system? Is the eDrilling system able to control last-generation rigs and handle technological conflicts? Is the eDrilling system able to tolerate different levels of expertise? Can eDrilling support the transition from classical drilling to remote drilling? Might the operator confuse the control cues used in the analogue task and those used in eDrilling?
	Safety
	To what extent are safety-critical eDrilling tasks exempt from the issue of cue change? To what extent can eDrilling assist in recovering control that has been lost due to missing cues or cue confusion?

Table 7.1   HMI drawbacks and corresponding questions *continued*

Drawback	eDrilling Assessment Questions
Opacity	HMI performance
	Is there drilling-related information that is needed by, but not available to the operator? Is there information available in the physical drilling workplace that should also be available in the eDrilling system but is not? Does eDrilling feed all needed information back to the operator? Can automation mask situations that can develop into problems? Can there be eDrilling decisions that might surprise the operator?
	Safety
	To what extent are safety-critical eDrilling tasks exempt from opacity issues? To what extent can eDrilling assist in recovering control that has been lost due to opacity issues?
Complexity	HMI performance
	Does the interface support the understanding of the internal and logical functioning of the eDrilling system? Are there interdependent control parameters in the eDrilling system? Can these interdependencies overwhelm the operator's processing capacity? Do system functions interact, or are there parameters that can cause control to be lost? Is the system so cognitively demanding that the operator can become complacent?
	Safety
	To what extent are safety-critical eDrilling tasks exempt from complexity issues? To what extent can eDrilling assist in recovering control that has been lost due to complexity issues?
Mode confusion and autonomy	HMI performance
	Does the interface support an understanding of the functioning of modes? Can modes change automatically? Are automatic mode changes announced to the operator? Is there potential for mode confusion on the part of the operator? Is full manual control available? Can the degree of autonomy/automation of eDrilling be adjusted by the operator? Can eDrilling take action without operator feedback? Can eDrilling override a human action without operator feedback?
	Safety
	To what extent are safety-critical eDrilling tasks exempt from mode confusion and autonomy issues? To what extent can eDrilling assist in recovering control that has been lost as a result of mode confusion or autonomy issues?

Table 7.1   HMI drawbacks and corresponding questions
               concluded

Drawback	eDrilling Assessment Questions
Anticipation	HMI performance
	Does the operator understand the eDrilling system sufficiently well to predict the behaviour of the system? Does eDrilling match or extend the operator's anticipation span? Is the data needed for anticipation easily accessible to the operator? Do the anticipation span and the type of data used in eDrilling accord with the operator's way of working?
	Safety
	To what extent are safety-critical eDrilling tasks exempt from anticipation issues? To what extent can eDrilling assist in recovering control that has been lost as a result of anticipation issues?

Assign Scores and Produce a Graphical Representation

How the assessment is scored depends on the questions that have been developed. For the purposes of this example, the following scoring system has been arbitrarily chosen:

- Questions assessing HMI performance can score between 0 and 2 (where 0 degrades HMI performance and 2 supports HMI performance).
- Questions assessing safety can score between 0 and 10 (where 0 is detrimental to safety and 10 is acceptably safe).

In this scenario, the scores for safety-related questions have been given greater weight. This reflects the emphasis that the prototype method places on safety. When the scores for individual questions are combined, each drawback is given a compound score. Table 7.2 takes Complexity as an example. Each drawback is processed in this way in order to determine a combined set of scores. This set of scores represents the eDrilling system's HMI performance and safety pedigree. The set of scores can then be displayed as a radar diagram (Figure 7.9) that highlights areas of strength and weakness.

Table 7.2   Example of assessment scores for Complexity

HMI Performance	Score
Does the interface support the understanding of the internal and logical functioning of the eDrilling system?	2
Are there interdependent control parameters in the eDrilling system?	1
Can these interdependencies overwhelm the operator's processing capacity?	2
Do system functions interact, or are there parameters that can cause control to be lost?	2
Is the system so cognitively demanding that the operator can become complacent?	2
TOTAL	9/10

Safety	Score
To what extent are safety-critical eDrilling tasks exempt from complexity issues?	9
To what extent can eDrilling assist in recovering control that has been lost due to complexity issues?	7
TOTAL	16/20
COMBINED TOTAL	**85/100**

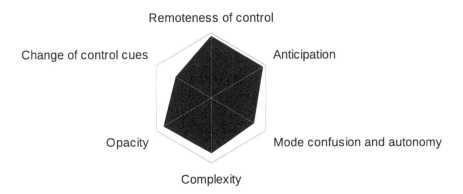

Figure 7.9   A radar diagram of fictitious eDrilling assessment scores

Assign scores to a risk class

The radar diagram can be used to assign the assessed system to a class of risks. This is done by deciding acceptability thresholds for each HMI dimension. For instance, the assessor can:

- set the lowest threshold for acceptability to 75, below which any single dimension causes the whole system to be rated as unacceptably risky;
- set a range between 76 and 90 where any single dimension causes the whole system to present acceptable risks; or
- set 91 as the upper threshold beyond which the assessed system is rated as safe.

These thresholds are not intended to determine what is acceptable or not. Rather, the aim is to show how decision criteria can be implemented into the method. This is an important feature of a risk assessment exercise where the assessment is usually only one step in the decision-making process.

Discussion

This chapter has described how the drawbacks of HMI can be turned into assessment criteria and incorporated into a prototype method for HMI performance evaluation. It is important to note that the proposed prototype method is only a demonstration of an instance of the set of criteria. It does not address issues such as what the set of criteria should include in terms of issues to be assessed, and how it should be used. These points (and several others) are discussed in the following section.

Assets

The prototype method proposed here is different from a traditional technically-centred risk assessment exercise as it assumes that a significant proportion of the performance of control and supervision systems is the result of HMI. This position differs from established theory in that it suggests that under certain conditions HMI can cause breakdowns in complex, dynamic, automated systems. Moreover, assessments of interactions between humans and machine interfaces typically only deal with the physical dimension of HMI (see for example Xiaoming et al., 2005). The approach taken here emphasizes that assessments must also target the cognitive dimension of the interaction.

The selected HMI assessment criteria and the method developed here can be applied to a wide range of interactive systems both within and beyond IO. The eDrilling scenario described here provided a test case, which demonstrated the deployment of a method that can be adapted to virtually any system that distributes control and supervision tasks between humans and machines.

Although function allocation has been mentioned several times in this chapter, the question of which functions to automate (or not), to what degree, and so on has not been addressed. Function allocation is a discipline in itself that spans many technical domains (Beevis et al., 1996). In the context of the general HMI assessment exercise it may require particular attention. For example, the assessment may look specifically at how functions are allocated between humans and machines, and how this allocation can be dynamically redistributed across tasks, under different time pressure settings, workload conditions and so on. More generally, any HMI evaluation should also include an assessment of how function allocation can support both performance and safety.

Limitations

This chapter does not discuss the design, development or integration of eDrilling. Concepts such as development-related risks (Boehm, 1991) or human–system integration (Booher, 2003, Pew and Mavor, 2007) are not addressed here.

It is also important to note that the generic automation-related factors must be 'translated' by a human factors specialist before they can be used as evaluation criteria. Namely, the cognitive implications of work on operators must be established and documented and these dimensions must be incorporated into the assessment questions. This can be done through a cognitive task analysis (for example) as recommended by Pierce and Salas (2003).

Finally, the assessment criteria discussed here do not include the ergonomic properties of the physical human–machine interface. Therefore, the work presented here must be supplemented by more traditional HMI considerations such as those described by Bastien and Scapin (1993), EU-OSHA (2009) or ISO 9241-110 (ISO 2006) (configurability, adaptability, error tolerance, ecological representativeness of the computerized task, operator training and so on).

Conclusion

While integrated drilling simulators are expected to be introduced into IO in the petroleum industry little is known about the risks created by the HMI. To address this question, this chapter reviewed a number of known drawbacks associated with the automation of systems, which provided the foundations for a prototype HMI assessment method.

The assessment dimensions discussed here appear to be applicable to domains other than the operational context of eDrilling. While it is difficult to list a comprehensive array of control and supervision tasks that these dimensions may be applied to, it seems reasonable to assume that they include (at a minimum) oil and gas-related activities such as control room operations and remote control and supervision tasks.

References

Abbott, K., Slotte, S.M. and Stimson, D.K. 1996. *The Interfaces Between Flightcrews and Modern Flight Deck Systems*. United States: Federal Aviation Administration.

AAIC (Air Accident Investigation Commission – Ministry of Transport). 1996. *Aircraft Accident Investigation Report 96-5, China Airlines, Airbus Industries A300B4-622R, B1816, Nagoya Airport, April 26, 1994.*

ACRC (Aeronautica Civil of the Republic of Colombia). 1996. *Controlled Flight into Terrain. American Airlines Flight 965 Boeing 757-233, N651AA near Cali, Colombia, December 20, 1995.* Accident investigaton report. Santafe de Bogota, DC, Colombia.

Albrechtsen, E. and Besnard, D. 2010. Preface. Views on Sociotechnical Vulnerabilities and Strategies of Control in Integrated Operations, in *Essays on Socio-technical Vulnerabilities and Strategies of Control in Integrated Operations* edited by E. Albrechtsen. SINTEF Report A14732 Trondheim, Norway: SINTEF-NTNU, 1–5.

Bainbridge, L. 1987. Ironies of Automation, in *New Technology and Human Error,* edited by J. Rasmussen, K. Duncan and J. Leplat. Chichester: John Wiley & Sons, 271–283.

Bastien, J.M.C. and Scapin, D.L. 1993. *Ergonomic Criteria for the Evaluation of Human-computer Interfaces.* INRIA, Technical Report No. 156.

Beevis, D., Essens, P. and Schuffel, H. 1996. *Improving Function Allocation for Integrated Systems Design: State-Of-The-Art Report.* (Cseriac Soar Series) Wright-Patterson Air Force Base, Ohio.

Besnard, D. and Baxter, G. 2006. Cognitive Conflicts in Dynamic Systems, in *Structure for Dependability: Computer-Based Systems from an Interdisciplinary Perspective,* edited by D. Besnard, C. Gacek, and C.B. Jones. London: Springer, 107–126.

Boehm, B. 1991. Software risk management: principles and practices, *IEEE Software,* 8(1), 32–41.

Booher, H.R. (ed.) 2003. *Handbook of Human-system Integration.* Hoboken, NJ: John Wiley & Sons.

Boy, G.A. 2005. *Human-Centred Automation of Transportation Systems.* AAET Conference, Braunschweig, Germany.

Crawford, E., Toft,Y., Kift, R. and Crawford C. 2010. Entering the Conceptual Age: Implications for Control Room Operators and Safety, in *Proceedings of the ICOCO 2010 International Control Room Design Conference.* Leicester: The Institute of Ergonomics and Human Factor.

Endsley, M. 1996. Automation and Situation Awareness, in *Automation and Human Performance: Theory and Applications,* edited by R. Parasuraman and M. Mouloua. Mahwah, NJ: Lawrence Erlbaum, 163–181.

EU-OSHA (European Agency for Safety and Health at Work) 2009. *Literature Review – The Human-machine Interface as an Emerging Risk.* Available at http://osha.europa.eu/en/publications/literature_reviews/HMI_emerging_risk [accessed 12 August 2001].

Fitts, P.M. 1951. *Human Engineering for an Effective Air/Navigation and Traffic Control System*. Washington, DC: National Research Council.

Ha, J.S. and Seong, P.H. 2009. A human-machine interface evaluation method: a difficulty evaluation method in information searching (DEMIS). *Reliability Engineering and System Safety*, 94(10), 1557–1567.

Hollnagel, E. and Woods, D.D. 1999. Cognitive systems engineering: new wine in new bottles. *International Journal of Human-Computer Studies*, 51, 339–356.

ISO (International Organization for Standardization). 2006. *NF EN ISO 9241-110. Ergonomics of Human-system Interaction. Part 110: Dialogue Principles*.

Meshkati, N. 2006. Safety and human factors considerations in control rooms of oil and gas pipeline systems: conceptual issues and practical observations. *International Journal of Occupational Safety and Ergonomics*, 12(1), 79–93.

METT (Ministère de l'Equipement, des Transports et du Tourisme; Ministry of Equipment, Transport and Tourism) (1993). Rapport de la Commission d'Enquête sur l'Accident survenu le 20 Janvier 1992 près du Mont Sainte-Odile (Bas Rhin) à l'Airbus A 320 immatriculé F-GGED exploité par la Compagnie Air Inter (Report of the Inquiry Commission on the Accident on 20 January 1992 near Mount Sainte-Odile (Bas Rhin) of the Airbus A320 registered F-GGED operated by Air Inter). Accessed on 31 May 2013 from http://www.bea-fr.org/docspa/1992/f-ed920120/htm/f-ed920120.html.

Parasuraman, R., Sheridan, T. and Wickens, D. 2000. A model for types and levels of human interaction with automation. *IEEE Transactions on Systems, Man and Cybernetics*, 30, 286–297.

Parasuraman, R. and Wickens, D. 2008. Humans: still vital after all these years of automation. *Human Factors*, 50(3), 511–520.

Pew, R.W. and Mavor, A.S. (eds) 2007. *Human-system Integration in the System Development Process: A New Look*. Washington, DC: The National Academies Press.

Pierce, L. and Salas, E. 2003. Linking Human Performance Principles to Design of Information Systems, in *Handbook of Human Systems Integration*, edited by H. Booher. John Wiley & Sons.

Rauterberg, M. 1996. How to Measure Cognitive Complexity in Human-computer Interaction, in *Proceedings of the Thirteenth European Meeting on Cybernetics and Systems Research* edited by R. Trappl. Vienna, Austria: Austrian Society for Cybernetic Studies, 815–820.

Rommetveit, R., Bjørkevoll, K.S., Halsey, G.W., Fjær, E., Ødegård, S.I., Herbert, M., Sandve, O. and Larsen, B. 2007. eDrilling: A System for Real-Time Drilling Simulation, 3D Visualization and Control, in *Proceedings of the SPE Digital Energy Conference and Exhibition*. Houston, TX: The Society of Petroleum Engineers.

Rommetveit, R., Bjørkevoll, K.S., Ødegård, S.I., Herbert, M., Sandve, O. and Halsey, G.W. 2008a. Automatic Real-Time Drilling Supervision, Simulation, 3D Visualization and Diagnosis on Ekofisk, in *Proceedings of the 2008 IADC/SPE Drilling Conference*. Orlando, FL: The Society of Petroleum Engineers.

Rommetveit, R., Bjørkevoll, K.S., Ødegård, S.I., Herbert, M., Halsey, G.W., Kluge, R. and Korsvold, T. 2008b. eDrilling used on Ekofisk for Real-Time Drilling Supervision, Simulation, 3D Visualization and Diagnosis, in *Proceedings of the 2008 SPE Intelligent Energy Conference and Exhibition*. Amsterdam, the Netherlands: The Society of Petroleum Engineers.

Sarter, N. and Woods, D.D. 1995. How in the world did we ever get into that mode? Mode error and awareness in situation Control. *Human Factors*, 37(1), 5–19.

Sheridan, T. and Verplank, W. 1978. *Human and Computer Control of Undersea Teleoperators*. Office of Naval Research, Technical Report N00014-77-C-0256.

Xiaoming, C., Zhiwei, Z., Zuying, G., Wei, W., Nakagawa, T. and Matsuo, S. 2005. Assessment of human-machine interface design for a Chinese nuclear power plant. *Reliability Engineering and System Safety*, 87(1), 37–44.

Chapter 8
Measuring Resilience in Integrated Planning

Kari Apneseth, Aud Marit Wahl and Erik Hollnagel

This chapter demonstrates how a Resilience Analysis Grid (RAG) can be used to profile the performance of a company in terms of the four abilities that characterize a resilient organization. It describes the development of a new, RAG-based tool founded on Resilience Engineering principles that can be used to assess an organization's resilience. The tool was tested in a case study involving a company in the offshore oil and gas industry. The company had decided to adopt an Integrated Operations (IO) approach to operations and maintenance planning and the tool was used to evaluate the impact of the Integrated Planning (IPL) process on its resilience.

The Four Abilities of Resilience Engineering

Resilience is defined as, '[…] the intrinsic ability of a system to adjust its functioning prior to, during or following changes and disturbances so that it can sustain required operations under both expected and unexpected conditions' (Hollnagel et al., 2010: 4). An organization is said to be resilient if it can adjust its functioning and respond appropriately to internal and external demands. Resilience Engineering focuses on four abilities that are essential for resilience, namely:

- *The ability to monitor*: monitoring organizational performance and identifying factors that both internally and externally can affect system performance.
- *The ability to respond*: responding appropriately to both regularities and irregularities by adjusting normal functioning.

- *The ability to anticipate*: knowing what to expect, that is, future developments, opportunities/threats, potential disruptions or changing operating conditions.
- *The ability to learn*: understanding what has happened, and learning the right lessons from the right experience.

The RAG is based on the four abilities outlined above; Hollnagel (2010) describes how it can be used as a tool for the continuous monitoring of organizational performance.

Integrated Planning

Traditionally, the planning of maintenance and operations in the offshore oil and gas industry has faced many challenges. Some of these include: the plurality of disciplines and activities involved, their different demands, unforeseen tasks, bad weather and logistical challenges related to material and personnel resources. The unpredictable environment found in offshore operations creates a situation where plans may have to be changed or abandoned at any moment. These constraints make it difficult for the organization to achieve optimal decision-making and responsiveness (Sleire and Wahl, 2008).

IPL attempts to provide a response to these problems. It implements IO principles and is designed to improve operational and maintenance planning in the oil and gas industry. The approach is holistic; the goal is to improve decision-making and create a more flexible and better-coordinated plan that can cope with the unpredictable nature of offshore operations (Ramstad et al., 2010). At the same time the IPL process aims to optimize the performance of the organization. It takes into account all of the task and resource demands of each of the various disciplines involved and integrates discipline-specific plans into one global plan. Tasks are evaluated and prioritized according to their criticality, in order to ensure that the right decisions (in terms of safety and production) are taken. The assessment also takes into account future constraints and any opportunities that can be identified. The result is a coordinated and feasible operational plan that can be executed with the necessary resources available.

Implementation of the IPL process requires several key enabling factors to be in place. The most important of these is Information and Communication Technology (ICT). These tools must provide real-time information and enable collaboration and communication across disciplines, organizations and geographical distance. Other important factors relate to organizational culture and include commitment, collaboration and competence. In addition, it is critical to establish organizational learning practices that include the use of Key Performance Indicators (KPIs) and continual adjustment processes (Ramstad et al., 2010).

Planning is organized according to various timescales distributed across a hierarchy of plans. Long-term planning concerns the realization of strategic goals and targets at a company level, which is expressed in strategic and tactical plans. Strategic plans describe 'what we want to accomplish', while tactical plans outline 'how we get there'. The long-term plan forms the basis for shorter-term planning (usually three months). In the short-term plan specific task schedules are defined and activities are described as work orders. The final detailed planning covers two-week periods. These provide a description of the offshore work that must be carried out in order to execute the plan, for each discipline involved. Incomplete tasks are added to a backlog and usually fed back into future plans.

Figure 8.1 shows this complex process. Production plans are shown on the left and logistics plans on the right. In order to establish an integrated plan, all of these individual plans must be incorporated into one single plan at all levels of the hierarchy. Coordination of the different plans must follow the continuous planning cycle shown in the centre of the figure. In this cycle, a new plan must be approved and communicated to all relevant parties; any changes to the plan are reported and provide input to future planning phases.

Resilience

As the pressure to increase production grows in the oil and gas industry, there is a need to improve efficiency while staying within the limits of safe performance. In the offshore environment, improvements in the resilience of the organization

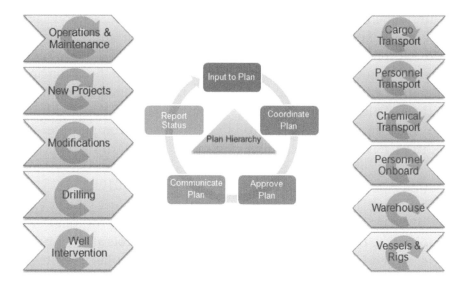

Figure 8.1   Production and logistics planning

Source: Centre for Integrated Operations in the Petroleum Industry, 2011

are expected to increase its flexibility and robustness. Moreover, resilience characteristics are thought to be an indication of the system's ability to cope with performance variability that is a natural part of socio-technical systems. Performance variability describes social and organizational factors found in the work environment, which affect the performance of complex systems. Variability can be found both in the external environment and in the various levels of the system itself (Hollnagel, 2009). It implies that unwanted incidents cannot be statically controlled simply through applying procedures and regulations to the organization. Instead, the system must develop properties that contribute to the production of safety.

Through focusing on the four resilience abilities, an organization can take a more dynamic approach to safety management. In the context of IPL the desired characteristics of a resilient system include: identification and coordination of critical functions; identification of future opportunities and constraints; continuous updating of plans (taking into account both programmed tasks and internal/external variability), and learning from both failure and success.

Method

This section describes the case study carried out in the spring of 2010. The company is one of the largest oil and gas operators on the Norwegian Continental Shelf and had decided to approach planning of operations and maintenance using IO concepts. The aim of the study was to develop a RAG-based tool to evaluate the IPL process and determine what impact the approach had on the organization's resilience.

The tool was developed in the following steps using a bottom-up approach:

1. *Identification of critical functions*: A literature review of IPL processes identified 16 critical IPL functions; these functions were sorted according to the four resilience abilities.
2. *Development of the tool*: The tool consisted of a checklist of questions based on the sixteen critical IPL critical functions identified in the previous step, and a measurement scale.
3. *Deployment*: Data was collected from the company's IO centre during a two-day period. Critical functions were identified and calibrated and the tool was tested. Interviewees included IPL process developers, staff involved in strategic and tactical planning and technical planners at an operational level. Two of the interviewees were asked to rate the company's IPL process based on the identified critical functions. The data was analysed using a theoretical Resilience Engineering framework.

Identification of critical functions

The first phase of the study established the critical functions of IPL. Critical functions are those processes that must function correctly in order to deliver outcomes. A literature review of IPL processes identified 16 critical IPL functions. These functions were sorted according to the four abilities of a resilient organization and are discussed below. Although presented separately it is important to keep in mind that all four resilience abilities are mutually dependent.

The ability to monitor

The monitoring ability of a company should enable short-term threats and opportunities to be identified (Hollnagel, 2010). This is an important part of any planning process, but even more so in a complex process such as IPL which requires an assessment of both factors that can threaten performance and factors that can create opportunities for the company. Monitoring is often carried out through KPIs, which are used to review system performance. In the IPL process, performance is defined as the ability to create an output that meets the intended purpose. Following are the four critical functions related to the monitoring ability.

1. External factors are monitored.
   - In the offshore environment, external factors such as the weather and resource shortages (human or material) can play a critical role in the execution of plans. These factors may delay planned activities: for example, bad weather can expose structures and equipment to heavy stresses requiring immediate maintenance. It is important to remember that external demands and constraints need not be physical; they can also include more intangible factors such as market pressure.
2. ICT tools can identify data sources.
   - To cope with performance variability, it is necessary to understand the system and its responses. ICT tools can be used to identify data sources. They help the user to understand system behaviour by clarifying how and when performance variability arises.
3. Changes in risk level are taken into account in planning.
   - The company must design a strategy that identifies the boundaries of safe performance and make decision-makers aware of these limits (Rasmussen and Svedung, 2000). The challenge is to translate this into actual IPL practice. Prioritizing short-term benefits at the expense of safety is likely to harm the organization in the long term. Short-sighted thinking, as well as decisions that are taken based on ambiguous criteria may gradually reduce safety margins in a distributed work system. Therefore it is important to develop clear, unambiguous decision-

making criteria for critical decisions (Rosness, 2001) and for the organization to define goals that prioritize safety.
4. ICT tools provide dynamic and up-to-date risk visualization.
   - Another challenge when designing complex systems is to develop criteria to measure constantly changing risk levels. ICT tools can provide a dynamic visualization of the current state of the asset. KPIs should provide up-to-date and dynamic information and should clearly communicate changing risk levels to the planning team; this in turn facilitates the planning process and improves decision-making.

The ability to anticipate
Understanding complex system events can be difficult, and anticipating future events is potentially even more difficult. However, it is important to prepare for the future, be proactive and adjust to any potential threats that may influence the system. In the preparation of action plans, future demands must be taken into account and activities that require scheduling and prioritising play a large part.

1. The performance measurement system anticipates changes in risk levels.
   - For optimal planning and prioritization of activities it is important to develop a realistic risk model that helps decision-makers reach the right decisions regarding the state of the asset. Tasks must be prioritized both individually and in a more complex risk scenario. As an extensive safety and risk analysis does not usually feature in the planning and prioritization of future maintenance activities, risks must be represented in a robust performance measurement system that can provide clear indicators of future changes in the risk level.
2. Opportunities, conflicts and execution problems can be anticipated.
   - The essence of planning lies in the anticipation of demands and it is critical to be able to react quickly when something unexpected occurs. There is a danger that planning can become a bureaucratic and static activity, therefore it is

important that the planning process is flexible enough to adapt to changing demands. A company's ability to plan successfully can in many ways be seen as an indication of how well it is able to anticipate the future in a constantly unstable context. As with monitoring abilities, event anticipation is about foreseeing both opportunities and threats. For successful planning, the internal environment (comprising the disciplines involved) must be understood.

3. Changes in the external environment can be anticipated.
   - The external environment may hamper internal coordination through its impact on the actual execution of plans. This means that external factors also need to be taken into account in planning. For example, seasonal changes in the weather may be relevant to some activities and seasonal changes in manpower capacity (for example, holidays) may be relevant to others. As oil and gas installations are located offshore, the execution of maintenance activities often relies heavily on the storage capacity of the facility. Otherwise, offshore crews must rely on good planning to provide them with the necessary resources at the appropriate time.

4. ICT tools provide a visualization of future demands.
   - It is vital to identify events and understand future demands, so that they can be taken into consideration when coordinating requirements. This can be enabled by the use of ICT tools that provide a visualization of future demands and that can communicate demands and opportunities.

The ability to respond

The response ability is closely related to the monitoring ability. The system must first identify that something has happened (monitoring ability), and then relate it to a category that the system has prepared a response for. IPL (together with preparatory activities, training, allocation of resources and so on) should function as a useful resource for the response and it should be possible to respond to both normal and unexpected events by an adjustment in functioning. In the context of IO the ability to respond is not defined as the actual execution of activities. Instead, response planning is about enabling the appropriate offshore response. Providing a timely and effective response is critical, and means that the measures that have been taken have had the desired outcome (Hollnagel et al., 2010).

1. Plans are continually updated.
   - In the context of IPL one important factor in the response is the ability to make continual updates (rather than at fixed intervals or after unexpected events) in order to ensure flexibility and highlight activity conflicts.
2. Planning is the responsibility of experienced personnel who understand activity conflicts.
   - A shared understanding of which activities have the highest priority at any given time is important in decision-making and in meeting the needs of the installation. A system that involves actors from many different disciplines also involves many decision-makers. In such a situation there is a high risk of sub-optimization as some decision-makers may simply negotiate for their own benefit. It is therefore important that decisions are made by experienced personnel who understand activity conflicts and who act on the basis of a common understanding of the goals to be achieved.
3. Activities can be easily reprioritized.
   - In the IPL process, the response must be appropriate to the aim of the process, and the purpose of scheduling is to avoid activity conflicts. But in the offshore environment, even excellent monitoring and anticipatory capabilities do not ensure that things always go as planned. It is necessary for the system to be flexible enough to make resources available upon demand and to be able to reprioritize activities in response to both internal and external events.
4. Short-term plans can be rescheduled.
   - As mentioned above, it is essential to have in place a plan or strategy for the continuous updating of the response capability. Specifically, it should be possible to reschedule short-term plans if a certain risk threshold is reached.

The ability to learn

In a multidisciplinary context there must be a strong focus on the quality of communication between the various actors involved, both in the initial stages of the planning process and as it develops. In order to cope with changes in a distributed working environment, the organization must have robust mechanisms for reporting, communication and learning.

1. Planning performance can be measured.
    *   In many planning activities, indicators are used (for example, 'plan attainment') to measure performance. However, in a dynamic planning process such as the IPL, these do not provide an accurate indication of planning quality. Hollnagel argues that safe systems are created through learning from both failure and success (Hollnagel et al., 2006). In other words, it makes more sense to understand the system through representative events rather than as a result of failure. In order to assess which events are representative, selection criteria need to be identified. These may include, for example, the identification of discipline-specific plans that must always be included in the integrated plan, and recurring activities that conflict, so that decisions can be taken efficiently and safely. This approach can be applied directly to the IPL process, as planning is, to a large extent, setting the right priorities while taking into account both opportunities and constraints. These points underline the importance of a robust measurement system able to measure the various dimensions of planning performance.
2. Knowledge is shared.
    *   The IPL process involves personnel from multiple disciplines who work together to optimize production. In this context, forward planning is challenging, as events that are considered critical by some might not be important to others and there is always the danger of bias in deciding priorities. Although it can be difficult to present individual agendas and formulate problems in such a way that they are comprehensible by everyone

involved, ICT tools and KPIs can provide a way to achieve this, and must be exploited. Key personnel should always be involved in developing and improving the planning process. Feedback can be helpful in making necessary adjustments and improvements. To enhance the planning process, other relevant personnel should interact and follow-up regularly.

3. Communication works both ways.
   - Offshore and onshore personnel may have different opinions on plan feasibility, and an exchange of views helps to improve the planning process. The feasibility of plans and improvements can be established through regular meetings between onshore and offshore teams, together with feedback from actions already taken and arguments from the planning team to justify their plans.

4. Improvement is continuous.
   - Learning includes the ability of the IPL process to continuously improve, correcting itself as potential conflicts emerge (for example, as work progresses or as a result of pilot tests). The IPL process can degrade when the organizational environment does not emphasize learning and communication.

The Resilience Analysis Grid-based tool

The previous section outlined how 16 critical functions of the IPL process were identified and related to each of the four resilience abilities. This section describes how these critical functions were operationalized. A tool was developed that consisted of a checklist of questions related to the 16 critical IPL functions, together with a scale that enabled each function to be rated.

The scale used to evaluate responses is that proposed by Hollnagel (2010), specifically:

> **Excellent (5):** the system on the whole exceeds the criteria addressed by the specific item.
> **Satisfactory (4):** the system fully meets all reasonable criteria addressed by the specific item.
> **Acceptable (3):** the system meets the nominal criteria addressed by the specific item.

Unacceptable (2): the system does not meet the nominal criteria addressed by the specific item.

Deficient (1): there is insufficient capability to meet the criteria addressed by the item.

Missing (0): there is no capability whatsoever to address the specific item.

The checklist itself was based on the 16 critical functions and contained sets of assertions grouped according to the four resilience abilities. The checklist is shown below in Table 8.1.

Table 8.1   Resilience analysis grid (RAG) checklist

	Excellent	Satisfactory	Acceptable	Unacceptable	Deficient	Missing
Monitor						
Short-term planning takes into account external factors that can lead to problems in the execution of activities.						
The ICT tools used in IPL make it easy to identify data sources.						
In the IPL process, changes in risk level are taken into consideration in planning future activities.						
ICT tools provide clear and up-to-date information and enable us to prioritize activities.						
Anticipate						
We have developed performance indicators that provide direct and indirect information on future changes in risk levels at our installations						
We anticipate opportunities, as well as conflicting tasks and execution problems through the use of our planning tools.						
Our medium and long-term planning takes into account external factors that may mean that planned activity will not be executed as expected.						
We have performant ICT tools that provide good visualization of future demands.						
Respond						
The integrated plan is continuously updated to take into account the changing needs of the installation.						

Table 8.1   Resilience analysis grid (RAG) checklist *concluded*

	Excellent	Satisfactory	Acceptable	Unacceptable	Deficient	Missing
Our planners are experienced in operations and understand the problems that may occur in the execution of activities.						
When problems in the execution of activities are identified, we can reprioritize and promote other activities.						
Active short-term plans are rescheduled when a certain risk threshold for the activities is reached.						
Learn						
We have a well-established performance measurement system to assess how well the IPL process is working.						
We emphasize shared experience and knowledge transfer among personnel working on IPL.						
Our planning includes well-established two-way communication between offshore and onshore teams.						
The IPL process is continually being improved.						

Deployment

The tool was evaluated and tested by personnel involved in the IPL process. The rating was performed by two key actors in the company's IPL; one was a developer involved in strategic and tactical planning, and the other was involved in planning at an operational level. They were asked to individually complete the checklist and rate the company's IPL performance.

The data collected was analysed using a Resilience Engineering theoretical framework in order to determine the impact the IPL process had had on the company's resilience abilities.

Results and Discussion

The ratings were shown in a similar diagram to the radar diagram below (Figure 8.2). The average score (from 0–5) for each question is shown on axes starting from the same point, and the set of 16 questions appear in the form of a star or radar. If possible, the scores should be normalized so that a regular polygon represents the intended state of affairs. However, even if that is not possible the star diagram clearly demonstrates irregularities in the scores, and also makes it possible to compare sets of scores collected over a period of time. Depending on the norms and expectations of the organization, a score of 'acceptable' or better may provide evidence that the organization's performance is sufficiently resilient.

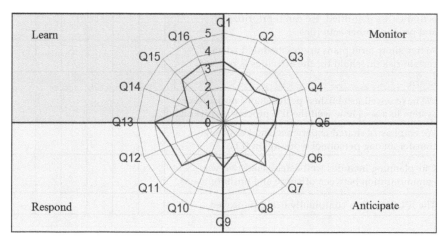

Figure 8.2   Sample RAG (not from the case study)

In the case study, only two ratings were made and it should be noted that the number of respondents must be taken into account when the tool is deployed. The results of the ratings were more or less consistent, albeit with minor differences. Both respondents rated all IPL functions as 3 (acceptable) or above, which demonstrated that the IPL process was enabling organizational resilience.

In some systems it may be acceptable for some abilities to have a higher criticality than others, but in a complex system such as the IPL process all four resilience abilities are equally important. A high level of performance in all areas is necessary for the system to function correctly.

In cases where the RAG indicates poor performance in some areas, a more thorough assessment should be undertaken to identify causes (that is, further study of the data collected, interviews, and examination of other relevant data sources) and improvement actions. The RAG questionnaire may provide useful input for assessments in cases where there are wide differences in ratings, or to identify which critical functions of the various abilities require attention.

Conclusion

An important point to highlight when using the RAG is that the relative importance of the four abilities varies depending on the domain in question. In practice, they are evaluated according to the needs of an organization. While all organizations must develop a response ability appropriate to their area of activity, and many organizations require a certain learning ability for them to improve their responses, not all organizations need the same emphasis on anticipating and monitoring abilities in order to be resilient. For example, a fire brigade needs a greater ability to respond to fires than it needs to anticipate fires (Hollnagel, 2010).

However, in the context of the high-risk, underspecified systems found in the oil and gas industry, it is clear that all four resilience abilities are required to be able to manage the uncertainties involved. In addition, there must be a stable reference point for managing, operating and developing the organization in such an unstable environment.

The case study showed, both through the interviews and ratings of company personnel, that IPL strengthened the organization's resilience through a vastly improved planning function. Through the introduction of flexible planning that could be changed if necessary, performance optimization that took into account the impact of the various disciplines on planning and execution of activities, and a clear focus on future changes and needs, IPL demonstrated how an organization can integrate resilience principles into both planning and daily operations.

An important consideration in using the RAG tool is how often measurements are taken. An organization experiences continuous change – in work processes, technical requirements, roles and responsibilities, and so on. This indicates that performing ratings need to be taken more frequently than once per year. A good starting point would be to make a fresh assessment each time a significant change is introduced into the organization. This would indicate whether new systems are functioning and represent an improvement to the system, or if new risks have been introduced. Through frequent measurements and following changes in results, the RAG-based tool can be used to monitor safety in any system.

This chapter discussed a stepwise approach for generating a resilience assessment in an IPL process. The method can be generalized and implemented in other systems and processes using the same steps, that is, the identification of critical functions and the development of a context-specific RAG based on the functions that have been identified. Once the critical functions have been operationalized and rated, the tool can function as a monitoring and performance management device for the optimization of system output and organizational performance.

Acknowledgements

This study was financed by the Center for Integrated Operations in the Petroleum Industry.

References

Centre for Integrated Operations in the Petroleum Industry. 2011. *The IPL Handbook.* Trondheim, Norway: MARINTEK.

Hollnagel, E. 2009. *The ETTO Principle: Efficiency-Thoroughness Trade-Off.* Farnham: Ashgate.

Hollnagel, E. 2010. RAG – the Resilience Analysis Grid, in *Resilience Engineering in Practice: A Guidebook*, edited by E. Hollnagel, J.Paries, D.D. Woods and J.Wreathall Farnham: Ashgate, 275–298.

Hollnagel, E., Tveiten, C.K. and Albrechtsen, E. 2010. *Resilience Engineering and Integrated Operations in the Petroleum Industry*, Report No. SINTEF A16331. Trondheim: Center for Integrated Operations in the Petroleum Industry.

Hollnagel, E., Woods, D.D. and Leveson, N. 2006. *Resilience Engineering: Concepts and Precepts.* Aldershot: Ashgate.

Ramstad, L., Halvorsen, K. and Wahl, A-M. 2010. Improved Coordination with Integrated Planning: Organisational Capabilities, in *Proceedings of the 2010 SPE Intelligent Energy Conference & Exhibition: Delivering Value – Creating Opportunities*. Utrecht: Society of Petroleum Engineers.

Rasmussen, J. and Svedung, I. 2000. *Proactive Risk Management in a Dynamic Society*. Karlstad, Sweden: Swedish Rescue Services Agency. Available at https://www.msb.se/RibData/Filer/pdf/16252.pdf [accessed 12 July 2012].

Rosness, R. 2001. *Om jeg hamrer eller hamres, like fullt så skal der jamres. Målkonflikter og sikkerhet* ['Whether I Pound or Am Being Pounded, There Will Still be Moaning!' Conflicting Objectives and Safety], SINTEF Report No. STF38 A01408. Trondheim, Norway: SINTEF. Available at http://risikoforsk.no/Publikasjoner/Ragnar%20STF38A01408.pdf [accessed 12 July 2012].

Sleire, H. and Wahl, A-M. 2008. *Integrated Planning, One Road to Reach Integrated Operations*. Presentation to the Society of Petroleum Engineers Bergen One Day Seminar. Available at http://bergen.spe.no/publish_files/2.3_Integrated_Planning_paper.pdf [accessed 2 May 2012].

Chapter 9

Resilient Planning of Modification Projects in High Risk Systems: The Implications of Using the Functional Resonance Analysis Method for Risk Assessments

Camilla Knudsen Tveiten

Resilience Engineering offers a new approach to safety management. The Resilience Engineering approach emphasizes the need to study normal operations of a socio-technical system to identify performance variability that may lead to unwanted consequences should interactions result in instability. The study discussed in this chapter implemented the Functional Resonance Analysis Method (FRAM) to identify potential performance variability in the planning of modifications to a mature offshore oil installation. In addition to identifying variability that may lead to unwanted outcomes in the planning phase of modifications, the study suggests that the FRAM may be useful as a risk assessment method at the organizational level.

Introduction

In recent years there has been an increased focus on a Resilience Engineering perspective on safety research (Hollnagel et al., 2006, 2008, Nemeth et al., 2009, Hollnagel et al., 2011). Earlier work by James Reason focused on the need to identify latent failures and highlighted latent errors as a hazard and a source of risk in maintenance activities (Reason, 1990, Reason and Hobbs, 2003). In the same vein, this chapter demonstrates the application of the Resilience Engineering approach as framework for managing emergent risks and latent failures.

A particular feature of the Resilience Engineering approach is that the study of normal operations can provide information about performance variability in the socio-technical system. Performance variability can lead to unwanted consequences when the variability of individual system functions combines in a way that destabilizes the system. The study described here aimed to identify performance variability that might be the source of risk during the planning of modifications on an offshore petroleum installation.

The planning of modifications to offshore oil and gas platforms is almost a daily activity in the industry, especially as installations mature. The planning phase in itself is not considered a high-risk activity, but modifications to operational installations can imply potentially radical changes to high-risk systems that may have a significant impact on safety. Complex production processes and sophisticated operating systems can make it difficult to identify risks. This particularly applies to older installations, as the older the installation is, the greater the likelihood that modifications will require the introduction of completely new technology. The consequence of modifications may be an intractable system that emerges either simply as a result of the modifications themselves, or in combination with other changes such as working practices or the organization of human resources.

The FRAM method has traditionally been used in accident investigations to model and analyse accidents in order to identify ways of reducing future risk. However, it has been argued that one of the strengths of the method, like other system-based accident models such as the Systems-Theoretic Accident Model and Processes (STAMP), is that it can also be used for risk assessment (Hollnagel, 2004, Leveson et al., 2006). FRAM has already been used as a risk assessment method in analyses at the operational level that aim to identify emergent risks due to the variability of normal performance (Macchi et al., 2009, Macchi, 2010). Similarly, the study described in this chapter investigates whether the FRAM method is useful for risk assessment in the planning phase of operations to identify potential performance variability; in this case, in the planning of modifications to a mature offshore oil installation.

The Functional Resonance Analysis Method

The FRAM method was first presented in the book *Barriers and Accident Prevention* (Hollnagel, 2004) and other comprehensive descriptions can be found in Macchi (2010) and Woltjer and Hollnagel (2007).

The basic element of the FRAM method is the function and its six descriptors (Input, Time, Control, Output, Resource and Precondition) as shown in Figure 9.1. A function is an action performed by a part of the system in order to fulfil the goals of the system. Functions can consist of both human activities and technological operations.

The FRAM model initially provides a simple visual representation of a system's individual functions. At this stage the connections between the functions of the model are potential rather than actual (that is, there is no attempt to evaluate function descriptors). It is only when the FRAM model is instantiated

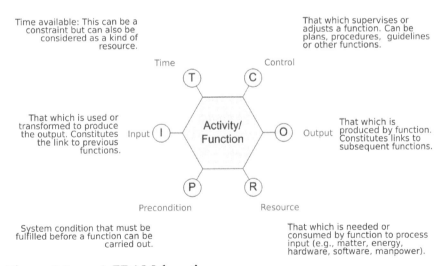

Figure 9.1   A FRAM function

Source: Hollnagel, 2008

that function descriptors are evaluated and the actual couplings between functions are established. Whether the FRAM method is used for accident or risk analysis, features of the operational environment can be added to the instantiation, which then provides a representation of the system's functions in a given time frame, conditions or scenario chosen for the purpose of the analysis.

A Functional Resonance Analysis Method analysis

Whether it is used for accident investigation or for risk assessment a FRAM event analysis in general consists of the following five steps:

1. *Define the purpose of the analysis.* The FRAM method can be used for safety assessment (analysis of future events) or for accident investigation (analysis of past events).
2. *Identify and describe the relevant system functions.* A function, in FRAM terms, is an activity or task which has important or necessary consequences for the state or properties of another activity. In the FRAM method, functions are identified based on 'work as done' rather than 'work as planned' (for example, procedures or work descriptions). Each function is characterized by six descriptors: Input (I, what the function uses or transforms), Output (O, what the function produces), Preconditions (P, conditions that must be fulfilled), Resources (R, what the function needs or consumes), Time (T, the time available), and Control (C, what supervises or adjusts the function).
3. *Assess the potential performance variability of each function.* The potential performance variability of a function is characterized qualitatively. Variability may originate in human intervention, the organization or technology, that is, the socio-technical system. The FRAM method distinguishes foreground and background functions, which may both be affected by variability. Foreground functions indicate what is being analysed or assessed, that is, the focus of the investigation and can vary significantly depending on the scenario, while background functions constitute the

context or working environment and vary more slowly. The variability of both foreground and background factors should be assessed as far as possible using either information provided by accident databases or on the basis of issues that are known to influence the behaviour of the system.

4. *Identify where functional resonance may emerge.* Functional resonance occurs when the normal variability of the system's functions interacts in unintended ways. Each function in the system exists in an environment composed of the other functions and every function has a normal variability. In certain circumstances, the variability of one function may act in such a way as to reinforce the variability of another (that is, resonate) and thereby cause variability to exceed normal limits. This phenomenon can be described as the resonance of the normal variability of functions, hence as *functional resonance*.

5. The fifth and last step is the *development of effective countermeasures*. These usually aim to dampen performance variability in order to maintain the system in a safe state, but they can also be used to sustain or amplify functional resonance that leads to desired or improved outcomes. Countermeasures are not discussed in detail here, but some measures to dampen variability are proposed.

The evaluation of performance variability

One of the general purposes of a risk assessment is to identify how and where incidents may arise. In the FRAM method, risk identification is based on an evaluation of the variability of normal performance. A review of case studies where a FRAM analysis has been applied shows that there are different ways of characterizing this variability. One approach was proposed by Hollnagel in the original description of the FRAM method (Hollnagel, 2004). Here, variability is evaluated by first identifying unexpected connections between system functions (for example, control loops or checks that may fail) and then evaluating how failure may occur (for example, the activity is not completed in time, lack of competence). The variability of each function and of the system as a whole is evaluated in the same way.

Another approach is to use common performance conditions (CPCs) to characterize the potential variability of a function. In this case, functions are evaluated as adequate, inadequate or unpredictable. CPCs were first introduced in the context of the Cognitive Reliability Error Analysis Method (CREAM) method (Hollnagel, 2007); they consist of 11 common conditions related to human performance. In more recent studies that have implemented the FRAM method, performance conditions are considered to be the outcome of other (typically organizational) functions (Herrera et al., 2010).

Another approach was proposed by Macchi et al. (2009) who suggested that system functions could be categorized into three types:

- *Human functions* are usually variable as people adjust their performance to current working conditions (resources and demands) (Hollnagel, 2009). Human performance can vary on a short-term basis, but may also have a dampening effect.
- *Technological functions* depend on the technology implemented in the system. These are less subject to variability as they are designed to be stable, reliable and predictable. Technical functions are not normally able to dampen performance variability (unless there are barriers in place in the system).
- *Organizational functions* are related to human functions but subject to a different kind of variability. Organizational functions are less variable than human functions – or rather their variability has a delayed effect on human functions. A typical example would be the production and updating of procedures.

The study then assessed the performance variability of the output of human functions using a three-by-three matrix that evaluated the temporal and precision characteristics of the function's inputs. Temporal characteristics were described as, 'too early, on-time and too late' and precision characteristics as, 'precise, acceptable and imprecise'. The result of this evaluation was an overall characterization of the function where the quality of the output was represented by the median of the quality of the input characteristics.

The Case Study

Oil and gas productions installations on the Norwegian Continental Shelf are owned by a licenced group of companies. One of the companies in the group holds the role of operating company. For maintenance and modification activities, work is often managed through an integrated service contract that is established between the operating company and an offshore construction company (the vendor company), which has the role of contractor and which may in turn contract other vendors. Integrated service contracts are long term and imply a broad concept of service that may require the contractor to undertake planning and administration responsibilities, provide cost estimates and so on.

Modifications to offshore oil installations can change the functionality of both a piece of equipment and the wider system. Updates to technology or the replacement of parts can make the system look and function differently. At the same time, drawings and manuals need to be updated. The goal is usually to improve the system or to adjust it to new requirements. Minor modifications can be carried out by the company's own staff or external contractors, and integrated into normal work processes. In this case decisions are taken locally within budgetary limitations. Major modifications however, involve the wider organization, including both the operating company and the contractor and may even require a temporary project team to be established.

At the installation in question, an integrated service contract had been in place since 2005. Modification projects followed procedures defined in manuals developed by the operating company and the construction contractor. Initially, each company had its own manual which was merged into a combined project plan detailing the roles and responsibilities of each of the two companies in different phases of the project. The planning phase consisted of: the identification of a problem or an opportunity; finding potential solutions; developing a modification procedure; preparing cost estimates; conducting a risk assessment; and pre-engineering activities. During this period the two parties communicated by telephone and email, and held formal meetings as prescribed by the project plan – typically when important decisions needed to be taken, or more generally to discuss and clarify ongoing work.

In this case planning activities involved the modification of an oil well. The problem was that production had fallen due to low well pressure. The proposed solution was to install a gas lift with subsea and topside components to increase well pressure. A gas lift involves injecting gas through a tubing-casing annulus. The injected gas aerates the fluid and reduces its density; the formation pressure is then able to lift the oil column and force the fluid out of the wellbore. This modification is one of several ways to increase oil well pressure and it is common in older wells. Nevertheless, the risk of a hydrocarbon leak during the installation and operation of the gas lift requires that a risk assessment be carried out. When the study began, the contractor had prepared a preliminary design for the topside construction of the lift. The modification was expected to be implemented during a planned maintenance period at the installation two months later. However, the work was postponed several times and still had not been carried out by the time the study ended.

Data gathering

Data for the FRAM analysis was gathered in two parallel phases. One phase consisted of an examination of both companies' manuals for the planning of modification projects. These manuals contain detailed descriptions of all the issues that must be taken into account during the planning phase, and procedures are described thoroughly and in detail as they must be generally applicable to all of the company's installations. Because of this level of generality, the functions that constituted the planning phase were clarified in discussions with project staff in both companies. The combined project plan that merged the two companies' work processes was also studied. The essential functions of the planning phase were extracted from these documents and recorded in a spreadsheet, together with a description of each function.

However, it is often the case that the detailed procedures prescribed in manuals describe the way work *should* be done; this frequently translates into a more manageable 'procedure' for day-to-day activities. This day-to-day 'procedure' is not described in any manual, but can be seen in the tasks, meetings and documentation that occur when modifications are planned. Therefore, the second data gathering phase consisted of observations of meetings between the operating company and the contractor, interviews with personnel in both companies and email exchanges with staff. The purpose of these observations and interviews was to validate that the analysis was based on the way planning activities were actually executed. It also provided the scenario that formed the basis for the instantiation of the FRAM model (the actual risk analysis). Field notes were taken of the observations and interviews for later use in the analysis. Particular care was taken to include variability in the way functions were executed.

Identification of system functions

Based on the information gathered from written procedures, interviews and observations, the functions of 'work as done' were identified. These functions are essential for completing the task when things go right in planning a modification project. Table 9.1 shows the functions that were identified, together with a description of their purpose and general characteristics.

These eight functions were described in terms of the six FRAM descriptors (Input, Output, Preconditions, Resources, Time and Control). For example, for the function 'define concept' the descriptions in Table 9.2 were prepared.

After describing all the functions according to their characteristics, a consistency check was performed. The consistency check aims to ensure that each descriptor is produced (and used) by at least one function in the model. The consistency check enabled the background functions to be identified. Background functions are essential for work to be carried out and include for example, the provision of resources, competence, budget planning and so on. The exact nature of background functions depends on the system being analysed. In this case, the background functions were identified through observations, interviews and procedures and are described in Table 9.3.

Table 9.1   Planning phase functions for modification projects

	Function	Description
1	Identify a problem or an opportunity	The need for a modification starts with either the identification of a problem or a proposal for improvement. A member of staff identifies a problem or an opportunity in the system that needs to be dealt with. This may include a suggestion for how to solve the problem or take advantage of the opportunity. *Human function* performed by the operating company.
2	Validate the proposal	The proposal is validated by a supervisor or manager before being entered into the database as authorized for further processing. *Human function* performed by the operating company.
3	Identify solution	Potential solutions for the problem or opportunity are investigated and one solution is chosen. Pre-estimates of cost and risk may be made. *Human function* performed by the operating company and the contractor.
4	Refine solution	Details such as the available resources, budget and criticality of the chosen solution are determined. *Human function* performed by the operating company and the contractor.
5	Define concept	The concept for carrying out the modification is described in detail. Any requirements and the impact on ongoing system operations are also described. The final decision on whether to make the investment is taken. *Human function* performed by the operating company and the contractor.
6	Estimate cost	Detailed costs estimates are prepared based on the description of the chosen solution and its impact on ongoing operations. This function applies to all phases of the modification project, with differences in the precision and level of detail. *Human function* performed by the engineering and finance departments using information provided by the contractor.
7	Assess risk	The risk assessment (economic, process-related and safety) is based on the description of the chosen solution and its impact on ongoing operations. Particular note is taken of the criticality of the work. This function applies to all planning phases. *Human function* performed by the operating company and the contractor.
8	Decision to proceed to pre-engineering	A kick-off meeting is arranged. The solution is opened for pre-engineering. *Human function* performed by the operating company.

These background functions resemble the organizational functions identified in the study by Macchi et al. (2009) who combined CPCs with research on the evaluation of safety-critical issues by Reiman and Oedewald (2009). However, in this case

Table 9.2   The function 'define concept' with descriptors

Descriptor	Function: Define concept
Input (I)	Defined solution selected Suggestion with criticality and possible solution registered in database
Output (O)	Concept study
Preconditions (P)	Solution selected Budget allocated
Resources (R)	Contractor personnel Engineering and maintenance personnel Expert technical personnel within the company
Time (T)	Possible solution available Criticality
Control (C)	Modification procedure

Table 9.3   Background functions

	Function	Description
1	Provide resources (operating company)	*Organizational function*: provide and manage sufficient human resources for adequate system functioning in the operating company.
2	Provide resources (contractor company)	*Organizational function*: provide and manage sufficient human resources for adequate system functioning in the contractor company (depends on contract).
3	Manage budget	*Organizational function*: provide and manage sufficient financial resources for adequate system functioning.
4	Manage modification procedure	*Organizational function*: design and update procedures to support activities.
5	Evaluate criticality	*Operational system function*: label the problem or opportunity with an attribute. For example, if the modification is assessed as critical for production continuity or safety, it is labelled as such based on pre-established criteria.

the organizational functions identified in the Macchi study were not appropriate to the analysis. For example, the authors identified a 'manage resources' function. In the current analysis this was divided into 'provide operating company resources' and 'provide contractor company resources' to account for the fact that resources had various origins, together with a 'manage budget' function to account for financial resources. In addition, the operational system function 'evaluate criticality' was added, as the criticality of a modification (in terms of safety and production issues) is an important part of the decision-making process that is not necessarily found in risk analyses.

The Functional Resonance Analysis Method

The foreground and background functions are shown in a FRAM model (Figure 9.2). Foreground functions are shown in grey and background functions in black.

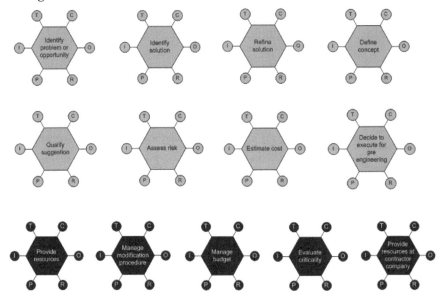

Figure 9.2   **The FRAM model for the planning phase of modifications**

This initial FRAM model shows all the *potential* couplings between functions, but not any *actual* couplings, which are the subject of the subsequent analysis.

Analysis of the normal scenario

The key to success for any modification project is that it is carried out on time and that cost and risk estimates are accurate. In addition, modification planning must take into account any other ongoing work on the installation, for example routine maintenance, other planned modifications, and normal production and operation activities. Modification projects form part of an overall work plan for the installation, which includes factors such as the availability of accommodation and transport (for personnel and equipment and so on).

The relationships between functions were based on the information provided by the function descriptors. The output of any function can be an input to another function. Taking the 'define concept' function as an example, inputs were 'defined solution selected' and 'suggestion with criticality and possible solution registered in database'. These in turn are outputs of the functions 'find solution' and 'identify problem or opportunity'. The fact that a solution exists (output from the 'find solution' function) is a precondition for the 'define concept' function.

Assessment of potential variability

Information about the potential variability of functions was gathered through observations of meetings and interviews. The description of potential variability is qualitative. In the analysis, potential variability in the output of each function was characterized as adequate, inadequate/inappropriate or unpredictable/missing. Where there was little performance variability and only small potential for interference from background functions, the output was characterized as 'adequate'. 'Inadequate' indicated that there was a greater potential for performance variability and interference from background functions, and when there was a lack of information on how the function was actually carried out, or the description of variability was unclear, the output was characterized as 'unpredictable'. The performance variability characteristics for the 'define concept' function are shown in Table 9.4. Similar tables were developed for all the functions in Table 9.1 and 9.3.

If any of the function descriptors I, P, C, R or T was rated as unpredictable or inadequate, the output (O) of the function was also assessed as unpredictable or inadequate, as unwanted variability in one function could transfer to any other function linked to it.

Table 9.4   Characterization of variability for the 'define concept' function

Descriptor	Function: Define Concept	Characterization
Input (I)	Defined solution selected Suggestion with criticality and possible solution registered in database	Adequate
Output (O)	Concept study	Inadequate
Preconditions (P)	Solution selected Budget allocated	Unpredictable
Resources (R)	Contractor personnel Engineering and maintenance personnel Expert technical personnel within the company	Adequate/ inadequate
Time (T)	Possible solution available Criticality	Unpredictable
Control (C)	Modification procedure	Adequate

Identification of functional resonance

The fourth step of the FRAM analysis aims to identify where functional resonance may emerge. This involves identifying the ways in which the variability in one or several functions may spread through the system. Particular combinations of variability may lead to situations where the system is unable to safely manage events.

The initial scenario instantiated in the FRAM model and described here illustrates how the FRAM method can be applied to a safety assessment of the planning phase; it could be followed by a further set of instantiations to represent the subsequent installation phase. Other instantiations could introduce features of the operational environment such as staff shortages, a lack of other resources or unforeseen problems with the well or production system.

In order to identify potential emergent risks in the planning scenario, two environmental features were introduced into the FRAM model: 1) multiple concurrent modification proposals and 2) an ageing installation (which normally increases the number of modification proposals). The analysis of the entire FRAM model is too complex to be described here. Instead, the selected extract (Figure 9.3) illustrates how the analysis is carried out and what the result may look like.

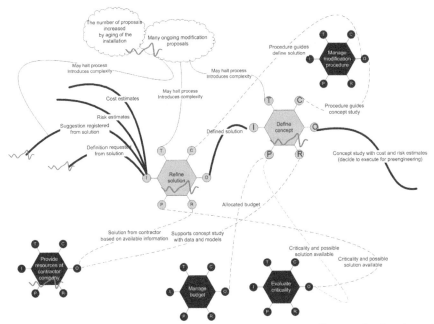

Figure 9.3 **Instantiation of the functions 'refine solution' and 'define concept'**

At the heart of Figure 9.3 are the two foreground functions 'define solution' and 'define concept' (shown in grey) and the relationships between them. In both foreground functions, the contractor is deeply involved in the preparation of proposals, preliminary concept descriptions and meetings and communication with the operating company. These functions also include preliminary risk assessments.

Figure 9.3 also shows background functions (in black) and the two environmental issues introduced to refine the instantiation (white clouds). Thick solid lines indicate the relationship between the output of one function and the input of another. Thick solid lines that are not connected to anything indicate an input or output to another function in the overall FRAM model (Figure 9.2) that is not shown here. Thin solid lines show relationships that are the result of the two environmental issues introduced for the purposes of the analysis. Thin dashed lines show the relationships between foreground functions and background functions. Finally, the zigzag lines indicate either variability in the function itself or the influence of variability from the input of another function.

The assessment of performance variability is based on the example shown in Table 9.4 and reveals several interesting results. Unwanted variability can be seen in both foreground functions. This problem relates to hidden or inaccessible information about the installation in question. Variability is strongly influenced by background functions that provide resources such as input from the contractor about similar modifications and information about the installation. In addition, both functions are influenced by the high number of modification proposals due to the age of the installation. Older installations that have undergone many modifications are typically underspecified and poorly documented. What documentation is available tends to be on paper and contained in drawings that have not been digitalized. Unwanted variability may also arise from the fact that the planning process follows detailed procedures that are remote from daily activities. 'Normal' work practices become habits that bear little relation to the specified procedures, and differing mental models of 'how' and 'how thoroughly' work needs to be done may develop in the organization. Another issue is that uncontrollable pressure in a gas lift in one well may impact other wells and the topside production system. This means that the entire production system needs to be taken into account in the risk assessment. Finally, other problems are created by the introduction of new employees who do not know the history of the installation. The combination of these factors can result in variability that may lead to unwanted outcomes in the planning phase. It is also likely to lead to delays which can put into question the decision to move into the pre-engineering phase.

It is important to note that the FRAM analysis helps to highlight both general variability in the process being modelled and where it is necessary to adjust to a specific situation. It is also important to focus not only on variability that may cause delays, adjustments, increased costs and so on, but also on variability that has an impact on safety. Moreover, it is important to ensure that the dampening of variability does not impede wanted variability and flexibility.

Figure 9.3 shows the instantiation of one part of the planning phase. In order to conduct a full risk assessment it is important to be able to zoom in and out of the overall model. In other words it should be possible to zoom in on a function and elaborate on the conditions that must be fulfilled in order to make it operable. It is also essential to be able to see the effects of functions on the whole system by zooming out of the model. This phase of the FRAM analysis depends on the description of functions in the model and emphasize the importance of ensuring that each function has been described in sufficient detail. This point is essential in order to arrive at an accurate risk assessment, to know where to dampen variability and to establish the relevant indicators necessary for monitoring (Herrera et al., 2010).

Countermeasures to dampen variability

In general, improvements to safety and efficiency in modification planning result from an examination of 'work as done' rather than 'work as planned'. Ideally, it should be possible to identify the functions that are necessary for the safe and efficient planning of modifications, and to use them as the basis for a discussion about improvements. In this respect, the development of the FRAM model may be a helpful countermeasure, as it highlights the fact that each function exists in the system, and all aspects of a function can be related to other aspects of other functions.

Another effective countermeasure may be to make information about the system and the installation (models, drawings, historical data and so on) visible and accessible to all personnel, and to encourage everyone involved in the planning and execution of modifications to work as a team, rather than two separate organizations. This is an issue that was addressed early on in the introduction of Integrated Operations (IO) in the petroleum industry (OLF, 2003) and should continue to be kept in mind. Modification planning would benefit from the sharing of updated and real-time information about the installation and its systems. In order for this to happen, barriers found in Information and Communication Technology (ICT) systems must be removed while at the same time ensuring safety.

An important concern is to dampen variability arising from the number of other modification projects and the fact that modifications may be carried out concurrently. It may be useful to develop a FRAM model specifically focused on this scenario. It may also be useful to group modifications on the installation (or to specific systems of the installation) into a single project or modification programme to provide a better overview and coordination of the situation.

Discussion

In order to use FRAM for risk assessment it must be possible to capture information about the human, technological and organizational functions of the system. A FRAM risk assessment requires that working practices are both planned and executed in real time and that information is available through observations and interviews. These factors suggest that FRAM is a promising risk assessment method for analysing the planning of high-risk modifications in an operational system.

As it is unusual to carry out a risk analysis of planning or other non-operational work processes it is difficult to evaluate whether the FRAM method provides a better insight into risks than other methods, as there is little comparable data. Nevertheless, accident investigations have demonstrated that the planning phase of a project is important in ensuring safe operations. As an example, in 2004 several deficiencies in the planning phase of a well operation at the oil production installation Snorre A resulted in one of the most serious near-accidents on the Norwegian Continental Shelf (Brattbakk et al., 2004, Schiefloe et al., 2005). Subsequent to the incident, the FRAM method was applied as a risk assessment method to the planning phase of the well operation (Phung, 2010).

Traditional risk assessment methods such as probabilistic risk assessment (PRA) characterize risks according to the magnitude (severity) of the possible adverse consequence(s), and the likelihood (probability) of occurrence of each consequence. As these methods rely on quantifiable issues and do not take account of issues that are not probable and/or have little impact, it is less likely that latent failures resulting from unwanted performance variability in the planning phase will be taken into consideration. These latent failures often combine with other issues and performance variability in the operational phase of the planned process, and eventually emerge as risks and hazardous situations. Major accidents can be the result of the combination of factors and behaviours that individually may have been regarded as quite normal and efficient if the result had been a success (Hollnagel, 2004, Hollnagel et al., 2006, Hollnagel, 2009).

The FRAM risk assessment may reveal dependencies between functions or tasks that are normally missed. In addition to providing recommendations for measures that may dampen variance (these measures reduce the chance of a negative outcome of an interaction or action in a system and may also be termed human, organizational or functional barriers) the FRAM method aims to provide recommendations for monitoring performance and variability, and detecting undesired variability. As a result of the implementation of the method, performance indicators can be developed for every function and every link between functions.

Although FRAM as a risk assessment method (and an accident investigation method) is still under development it has been suggested that it may be a promising way to identify safety indicators (Herrera et al., 2010). As an accident investigation method, FRAM has already been applied to accidents and incidents in aviation (Woltjer and Hollnagel, 2007) and other settings. Experience has shown that FRAM is well-suited to the investigation of system accidents and that it provides a better understanding of events than traditional multi-linear methods (Herrera and Woltjer, 2009).

Conclusion

The study described here suggests that the FRAM method is well-suited to the investigation of complex socio-technical operations and work phases in the oil and gas industry. This conclusion is based on the fact that traditional risk assessments rarely assess risk in normal working conditions (such as planning), although accident investigations have demonstrated that socio-technical factors are very often found to be contributing factors to incidents.

This study also demonstrates that further work is needed if the FRAM method is to become a practical tool for risk assessment in the offshore oil and gas industry. This point particularly applies to the evaluation of potential variability and how this should be assessed. It is also important that the method is made accessible to risk analysts and safety personnel in companies; although it will probably be necessary to provide research support for the analyses while the method is under development.

In addition, efforts should be made not only to dampen unwanted variability in the system, but also to support desirable performance variability. In the study in question, this could take the form of making historical information about the installation available to everyone, and facilitating the exchange of real-time data between the operator and the contractor company. Organizational barriers should be lowered in order to provide support for the integration of employees who share the same goal; safe and efficient modifications. Although further development of the FRAM method is needed in order for it to become of practical use in a dynamic operating company, the method shows promise as a tool for the identification of latent failures and potential variability in the planning phase of high-risk operations.

Acknowledgments

The study described here was financed by the Center for Integrated Operations in the Petroleum Industry.

References

Brattbakk, M., Østvold, L.-Ø., Van der Zwaag, C. and Hiim, H. 2004. *Investigation of the Gas Blowout on Snorre A, Well 34/7-P31A, 28 November 2004*. Stavanger: Petroleum Safety Authority. Available at http://tinyurl.com/d67bgqz [accessed 16 July 2012].

Herrera, I.A., Macchi, L. and Hollnagel, E. 2010. *What to Look For When Nothing Goes Wrong? A systematic Approach to Monitor Performance Variability*. Unpublished work.

Herrera, I.A. and Woltjer, R. 2009. Comparing a Multi-Linear (STEP) and Systemic (FRAM) Method for Accident Analysis, in *Safety, Reliability and Risk Analysis. Theory, Methods and Applications,* edited by S. Martorell, C. G. Soares and J. Barnett. CRC Press, Taylor and Francis Group, 1: 19–27.

Hollnagel, E. 2004. *Barriers and Accident Prevention*. Aldershot: Ashgate.

Hollnagel, E. 2007. *CREAM (Cognitive Reliability and Error Analysis Method)*. Available at http://tinyurl.com/bm9x3aa [accessed 16 July 2012].

Hollnagel, E. 2008. *From FRAM (Functional Accident Resonance Model) to FRAM (Functional Accident Resonance Method.* Presentation at the 2nd FRAM Workshop, Sophia Antipolis, Ecole des Mines, Centre for Research on Risks and Crises.

Hollnagel, E. 2009. *The ETTO Principle: Efficiency-Thoroughness Trade-Off . Why Things That Go Right Sometimes Go Wrong.* Aldershot: Ashgate.

Hollnagel, E., Nemeth, C.P. and Dekker, S. 2008. *Remaining Sensitive to the Possibility of Failure.* Resilience Engineering Perspectives, Vol .1. Aldershot: Ashgate.

Hollnagel, E., Paries, J., Woods, D. and Wreathall, J. (eds) 2011. *Resilience Engineering in Practice: A Guidebook.* Ashgate Studies in Resilience Engineering. Burlington, VT: Ashgate.

Hollnagel, E., Woods, D.D. and Leveson, N. (eds) 2006. *Resilience Engineering: Concepts and Precepts.* Aldershot: Ashgate.

Leveson, N., Dulac, N., Zipkin, D., Cuther-Gershenfeld, J., Carroll, J. and Barrett, B. 2006. Engineering Resilience into Safety-Critical Systems, in *Resilience Engineering: Concepts and Precepts* edited by E. Hollnagel, D. Woods and N. Leveson. Burlington, VT: Ashgate, 95–124

Macchi, L. 2010. *A Resilience Engineering Approach to the Evaluation of Performance Variability: Development and Application of the Functional Resonance Analysis Method for Air Traffic Management Safety Assessment.* PhD thesis. France: Ecole Nationale Supérieure des Mines de Paris.

Macchi, L., Hollnagel, E. and Leonhardt, J. 2009. *Resilience Engineering Approach to Safety Assessment: An Application of FRAM for the MSAW System,* EUROCONTROL Safety R&D Seminar. Munich, Germany. Available at http://tinyurl.com/cb3v8bl [accessed 16 July 2012].

Nemeth, C.P., Hollnagel, E. and Dekker, S. 2009. *Resilience Engineering Perspectives: Preparation and Restoration.* Aldershot: Ashgate.

OLF (Norwegian Oil Industry Association). 2003. *eDrift på norsk sokkel – det tredje effektivitetspranget* [eOperations on the Norwegian Continental Shelf – the Third Efficiency Leap]. Available at http://tinyurl.com/cvecjtc [accessed 16 July 2012].

Phung, V.Q. 2010. *Bruk av functional resonance analysis method (FRAM) som et verktøy for å forutse og identifisere fremtidige uønskede hendelser* [Using the Functional Resonance Analysis Method to Anticipate and Identify Future Incidents]. Master's thesis. Trondheim: NTNU.

Reason, J. 1990. *Human Error*. Cambridge: Cambridge Press.

Reason, J. and Hobbs, A. 2003. *Managing Maintenance Error. A Practical Guide*. Aldershot: Ashgate.

Reiman, T., and Oedewald, P. 2009. *Evaluating Safety-critical Organizations – Emphasis on the Nuclear Industry*. Swedish Radiation Safety Authority. Available at http://www.vtt.fi/inf/julkaisut/muut/2009/SSM-Rapport-2009-12.pdf [accessed 30 April 2012].

Schiefloe, P. M., Vikland, K. M., Torsteinsbø, A., Ytredal, E. B., Moldskred, I. O., Heggen, S., Sleire, D. H., Førsund, S. A. and Syvertsen, J. E. 2005. Årsaksanalyse etter Snorre A hendelsen 28.11.2004 Stavanger [Causal Analysis of the Snorre A 28.11.2004 Stavanger Event]. Statoil ASA.

Woltjer, R. and Hollnagel, E. 2007. The Alaska Airlines Flight 261 Accident: A Systemic Analysis of Functional Resonance, in *Proceedings of the 14th International Symposium on Aviation Psychology*. Dayton, OH.

Chapter 10

Promoting Safer Decisions in Future Collaboration Environments – Mapping of Information and Knowledge onto a Shared Surface to Improve Onshore Planner's Hazard Identification

Grete Rindahl, Ann Britt Skjerve, Sizarta Sarsha and Alf Ove Braseth

This chapter discusses a study that investigated how shared collaboration surfaces may improve hazard identification in maintenance planning involving distributed teams on the Norwegian Continental Shelf. It describes a software prototype designed to present risk information. Evaluation studies of the prototype show that the tool is perceived as valuable by users.

The introduction of Integrated Operations (IO) in the petroleum sector on the Norwegian Continental Shelf has raised a number of questions about the implementation of innovative operational technologies intended to enable these new working practices (Holst and Nystad, 2007, OLF, 2008). It cannot simply be assumed that new technologies will improve collaboration between teams distributed across the Norwegian Continental Shelf and in certain situations they may even complicate task performance; for example, they may make it more difficult to establish shared situation awareness and trust between team members. Moreover, in a context where Health, Safety and Environment (HSE) issues permeate all decision-making processes, another key issue is whether these new technologies and working practices can improve an organization's ability to identify and mitigate hazards. In practice, the extent to which new technology enables a team or an individual to correctly perform their tasks will vary, depending on how well the technology fits the needs of its users.

According to the guidelines of the Norwegian Oil Industry Association (OLF, 2011) all personnel involved in the planning, approval and execution of work must consider whether a work permit is needed for the work in question. Furthermore, all personnel involved in the task must consider whether a safe job analysis should be performed. Permission to work is issued for each defined work operation at a given location under given prerequisites and with the requirement that it be executed in a safe manner.

While long-term plans are prepared by onshore planners, maintenance planning is mainly performed offshore. Currently, the initial planning period (dating back days or often weeks) does not necessarily include a risk assessment of a specific task or a combination of tasks, and hazards are only assessed during the offshore period of task planning; usually starting 24 hours before a set of tasks are to be executed. This fact, combined with limited capacity and last-minute corrective maintenance means that activities frequently have to be re-planned. Consequently, each day a large number of tasks are sent back for re-planning, and another set of tasks is scheduled for execution.

Method

The study was divided into two phases:

1. A pre-study that examined the characteristics of current working practices and technological bottlenecks in distributed teams operating in an IO context. This was achieved through a review of structured observations of collaboration and a series of visits to the industry.

2. A usability test of a shared collaboration surface developed for the identification of hazards related to task planning.

The pre-study: Identification of system characteristics

Decision-making in a distributed team environment can be challenging. Two earlier structured observational studies of IO collaboration identified the characteristics of shared collaboration surfaces that may affect decision-making in distributed teams (Rindahl et al., 2009, Kaarstad et al., 2009).The Rindahl study in particular demonstrated that highly skilled team members were able to use shared surface technology to improve and focus collaboration:

> As an example, HSE received the desired attention because it was first on the agenda and had enough information both on the registered minor incidents or reported good HSE, as well as on the special HSE focus of each discipline for the day. The same amount of space that HSE was given on this surface it also seemed to get in the meeting.
>
> Further, with a highly technologically literate meeting leader, pointing and changing the surfaces could be used as a means to keep the discussion focussed and efficient, to make people take the same leaps together and to summarise and conclude. It is not likely that the same focus would have been achieved without this shared surface channelling everybody's attention in the same direction... (Rindahl et al., 2009: 32).

These findings were complemented by field visits and surveys in six different organizations. The results of this pre-study showed that distributed teams required collaborative work surfaces that made it easier to share information and knowledge. Specifically, the characteristics they needed included the following:

*   a *holistic* visualization of current status and plans;
*   to be able to visualize the *experiences and ideas* of the meeting's participants; and
*   to be able to visualize the meeting's *consensus and actions*. This was crucial both for efficient collaboration and for maintaining compliance with decisions and actions (Rindahl et al., 2009).

Another key factor was that risks:

- could be identified early in the planning process, to reduce the need for re-planning;
- could be easily discussed and communicated across disciplines; and
- could be mitigated, through a collaborative approach where team members provide complementary knowledge and competencies.

A further important point that emerged from the pre-study was that the lessons learned from earlier decisions were not always easily incorporated into maintenance management systems. Several companies pointed out that increasing their ability to store and retrieve previous decisions and lessons learned would be a great improvement and would enable faster and safer decisions. Furthermore, improved knowledge transfer of past decisions would provide a better overview of the implications of upcoming decisions. These comments were supported by an examination of a cross-section of investigation reports (for example, from the Petroleum Safety Authority Norway website) which showed that factors such as working practices, operating procedures, staff training and maintenance were repeatedly found to contribute to accidents.

These findings highlight the fact that, in order for shared collaboration surfaces to contribute to improvements in an organization's ability to identify and mitigate risk, they need to have a high level of usability.

Usability testing

The software test bed

The software test bed, called the IO Maintenance and modification Planner (IO-MAP) was designed to contribute to safer decision-making in future collaborative environments (that is, future IO work practices) (Sarshar et al., 2010). It maps risk information and knowledge related to planned maintenance activities onto a collaborative surface. A collaborative surface is a large, flat, graphical display that can be monitored and potentially operated by all members of a distributed team, regardless of their physical location.

Information is defined here as data that '(1) has been verified to be accurate and timely, (2) is specific and organized for a purpose, (3) is presented within a context that gives it meaning and relevance, and (4) that can lead to an increase in understanding and decrease in uncertainty'. And knowledge is defined as, '1. General: Human faculty resulting from interpreted information; understanding that germinates from combination of data, information, experience, and individual interpretation.'[1]

The IO-MAP was inspired by Geographical Information Systems (GIS) technology and software such as VR Dose – a tool used in the nuclear industry for the visualization of virtual reality radiation systems (Rindahl et al., 2009). These tools help in the understanding of risk by providing a visual representation of the working environment. Task conditions are shown on a flat, map-like interface, and users can clearly see tasks that have been scheduled in the same time period and/or in nearby locations. Furthermore, the interface makes it possible to visualize workloads and job combinations within areas and disciplines, and presents them using a timeline. The software is based on goal conflict theory (Skjerve et al., 2009) and is designed to visualize and present two overall types of information: 1) hazards, and 2) standards/rules to determine how these hazards should be handled. Although the IO-MAP can be used to support the planning and modification of maintenance activities, it is first and foremost a tool intended to improve the overall understanding of hazards in the planning process and to enable their early identification. One way that the IO-MAP achieves this is to enable planners to identify or raise questions about hazards at an early stage.

The collaborative surface was developed on the basis of the findings of the pre-survey; the aim was to provide an easily understandable overview of notifications, work orders and work permits together with information on work in various stages of

---

1 See http://www.businessdictionary.com/definition/information.html and http://www.businessdictionary.com/definition/knowledge.html.

planning and the known conditions of the work area. Furthermore, the user(s) was able to add their own knowledge and observations to the surface, thereby sharing their views and understanding of risk with others, either in a collaborative setting or individually at a later point in time. A screenshot of the prototype implemented for the first usability study is shown in Figure 10.1.

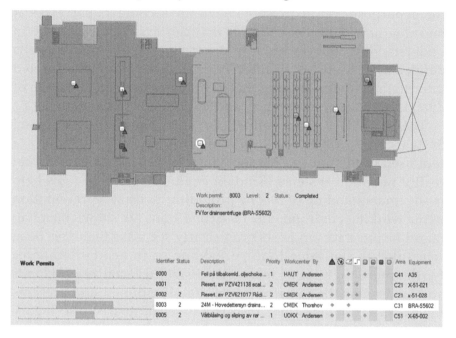

Figure 10.1   Screenshot from IO-MAP as used in the first usability study

The design rationale used in the first version of the IO-MAP display (Figure 10.1) was inspired by the Information Rich Display design scheme developed for process control in the oil and gas industry (Braseth et al., 2004), which was later used in nuclear applications (Braseth et al., 2009, 2010). The Information Rich Display design scheme aims to present a high density of data while remaining readable. This is essential in highly complex environments where it is important to avoid keyhole effects (that is, the display only reveals a fraction of the total process) and problems for the operator who may be forced to navigate extensively through the system to obtain the required risk-related information.

The colour principles used in the Information Rich Display design concept were inspired by Tufte (Tufte, 1983, 1990, 1997). The colour concept uses a visual layering technique whereby the most important information is presented visually on the most salient colour layer, for example, red/yellow. Less important information is presented using less visually salient colours such as grey. Frames, lines and shading are avoided in order to keep the display from looking cluttered and crowded. These techniques make it possible to create an information rich screen while keeping the display readable. Design techniques (such as the alignment of data in columns and rows) were applied to further enhance the display. The display also implements pattern recognition techniques (Rasmussen, 1983), which aim to reduce the cognitive stress found in traditional designs. It uses a stable information frame, without pop-up windows or hierarchical information presentation. This enables the end-user to recognize information not only by its text or symbol, but also by the specific position of the data.

Determining the appropriate level of automation in a complex display design is challenging. The information presented in the IO-MAP test bed display is multivariable by nature. The safety variables that can be identified in a project range from work type to noise, explosion and platform zones (it should be noted that this is only the first prototype of the display, and later versions will introduce further safety-related variables). Moreover, safety variables are inherently different from the automated variables found in the process industry (for example, temperature, flow and pressure). These systems can be automated with the aid of simple Proportional–Integral–Derivative (PID) controllers, and predefined limits that alert the operator when an out-of-range situation occurs. Instead, the aim of the IO-MAP display design is to present data in a way that supports a type of cognitive augmentation. Safety-related data are presented to the end user in a way that improves their understanding of the total risk and hazard. Alarms and pop-ups that may lead to a mistaken or incomplete understanding of a complex situation are not implemented.

Participants and testing
A usability test of the IO-MAP interface (Skjerve et al., 2011) was set up to gather data on the ability of planners to identify HSE-related

hazards when working with the shared surface. Both subjective data (users' impressions of the tool and its usability) and performance data (the users' ability to solve problems with the tool) were collected. Participants were given an operational plan and a set of tasks described in the IO-MAP. Several hazards ('Easter eggs') were hidden in the task combinations. Participants were asked to use the tool to review, comment, reprioritize and re-order the given tasks in order to draw up an appropriate plan. Data was collected through observations, interviews, a questionnaire and system logs.

Eight representative users participated in the study. The eight users came from four petroleum companies: Statoil, Eni Norge, Shell and GdF Suez. They were selected by their employers from a pool of potential future users of the IO-MAP (Skjerve et al., 2011). Each user participated individually.

Scenarios
Scenarios were developed in close collaboration with Statoil personnel. A Statoil asset was used as the example in the tool, and scenarios were based on real data from their maintenance management system, together with HSE profiles and other platform data. Each member of the project team held a specialized function – they included a platform manager, the onshore HSE coordinator, an onshore planner with extensive offshore experience and a representative from DNV (Det Norske Veritas, an independent risk management foundation) with a strong risk focus and extensive field experience. This process of creating scenarios also provided input to, and an evaluation of, the IO-MAP itself.

Users were presented with an IO-MAP description of a day, consisting of scheduled work orders and work permits on the example asset. The proposed plan far exceeded normal conditions both in terms of workload and hazards. They were then told to correct the plan by moving tasks in the timeline so that risk levels would remain acceptable. Furthermore, they were asked to add information about the hazards they perceived in the system.

Data collection
Data was collected through a questionnaire, interviews, observations, expert ratings and system logs. The Hazard

Perception questionnaire consisted of the following three statements about hazard perception, related to the type of hazards presented in the IO-MAP:

1. In general, the IO-MAP showed me hazards before I identified them myself.
2. In general, the IO-MAP showed me rules/standards for handling hazards before I thought of them myself.
3. Overall, the IO-MAP provided very good support in developing plans in which safety issues were adequately dealt with.

Users responded on a seven-point, ordered, one-dimensional scale.

Data was also obtained from a subset of semi-structured interviews, which addressed issues related to risk perception. In addition, observations were collected during the usability test.

Results

Overall, the study showed that the support provided by IO-MAP in hazard visualization was perceived as valuable by users. The average score from the Hazard Perception questionnaire on both the first and second questions was 5.2 (out of a maximum of 7). This indicated that the IO-MAP helped to visualize hazards and provided rules for how these should be handled, before the user was able to do so on their own. On the third question (overall evaluation) the average score was 5.6, indicating that the users were highly satisfied with the way the IO-MAP presented hazards.

The outcome of interviews supported the above interpretation. These revealed that the IO-MAP was perceived as improving safety, by helping the user to identify and remain focused on hazards. Overall, users found that the representation of risk was helpful (task-related risks were shown by a yellow triangle, and risks and hazards associated with combinations of tasks were shown by black lines). Most users also appreciated the depiction of zones (for example welding/no-welding), a noise chart and the ability to freely and easily enter important (or potentially important) comments in order to improve risk visualization. This included the addition of notifications (a new job that needs to be included in the planning), potential threats (users could enter hazards and/or concerns related to risk based on their own perception of the situation), work permits and final task planning. In general, users assessed that the IO-MAP contributed to improved communication between team members with different backgrounds, as it made it possible to clearly visualize the various potential elements of the plan.

At the individual level, the IO-MAP was said to have made it easier for the user to immediately recognize hazards (for example, combinations of tasks). Specifically, one user mentioned that the potential hazards included in the display of predefined work types were something that onshore planners would not normally pay attention to. Currently, this type of information is a matter for offshore staff, when work permits are printed. Generally, this user found that the IO-MAP helped them identify hazards earlier than they would normally have done. Another user stated that the displayed map interface served as a visual reminder of the hazards associated with the situation in question. A third user, who had worked offshore, found it useful as a way to transfer a better understanding of the offshore environment to onshore colleagues with more limited offshore experience. This statement was supported by others with limited offshore experience, who stated that the IO-MAP improved their understanding of offshore work, as it showed where planned tasks should be performed. Another comment was that the IO-MAP offered a much better depiction of active work permits than the existing maintenance management system.

Planners who had previously worked offshore (often those with pre-IO experience) demonstrated an advantage in identifying hazards at an early stage. This is likely due to their offshore knowledge and gut-feeling. However, indications are that the visualizations provided by the IO-MAP may help planners who never worked offshore to understand others' experience and develop the same gut feeling.

Initial findings suggest that the tool may be deployed across disciplines, albeit with slightly different interface settings.

However, terminology maybe a challenging factor, and this is under further investigation. Feedback from the users participating in the study indicated that the technology could be turned into groupware for distributed collaboration, with some additions to accommodate the differing needs of users. The usability study confirmed the pre-survey finding that better keeping of, and access to, the results of previous decisions was needed. Record-keeping and traceability will therefore remain an important factor in future work. A range of suggestions were put forward to improve the visualization. These included the addition of weather data, the risk of falling objects and areas affected by crane operations.

Conclusion

The study described here attempted to provide some answers to questions such as: How can new technologies improve collaboration in distributed teams on the Norwegian Continental Shelf? To what extent is technology an obstacle rather than an asset? What is needed for technology to improve decision-making? These challenging questions continue to motivate our research, which seeks to identify potential problems and provide solutions before new technology is implemented. This chapter looked at one particular type of new technology, the shared collaborative surface, and investigated the characteristics of the information and knowledge that should be shown on the shared surface in order to facilitate risk-informed decisions in future collaboration environments. While the initial results are promising, the prototype IO-MAP focuses on IO in future settings and many of the issues it addresses are still emerging.

Acknowledgements

The authors wish to thank the Center for Integrated Operations in the Petroleum Industry and all partners involved in this research (Statoil, DNV, GdF SUEZ E&P Norge, Norske Shell, Eni Norge). We also wish to thank our colleagues in the IFE FuCE and SOFIO teams.

References

Braseth, A.O., Karlsson, T. and Jokstad, H. 2010. Improving Alarm Visualization and Consistency for a BWR Large Screen Using the Information Rich Concept, in *Proceedings of the 7th ANS Topical Meeting on Nuclear Plant Instrumentation, Controls, and Human Machine Interface Technologies*. Las Vegas, NV: American Nuclear Society.

Braseth, A.O., Nurmilaukas, J. and Laarni, J. 2009. Realizing the Information Rich Design for the Loviisa Nuclear Power Plant, in *Proceedings of the 6th ANS International Topical Meeting on Nuclear Plant Instrumentation, Controls, and Human Machine Interface Technologies*. Knoxville, TN: American Nuclear Society.

Braseth, A.O., Veland, Ø. and Welch, R. 2004. Information Rich Display Design, in *Proceedings of the 4th ANS International Topical Meeting on Nuclear Plant Instrumentation, Control and Human Machine Interface Technology*. Columbus, OH: American Nuclear Society.

Holst, B. and Nystad, E. 2007. Oil and Gas Offshore/Onshore Integrated Operations – Introducing the Brage 2010+ Project, in *Official Proceedings of the Joint Meeting and Conference of the Institute of Electrical and Electronics Engineers (IEEE) and Human Performance/Root Cause/Trending/Operating Experience/Self-Assessment* (HPRCT). Monterey CA: HRPCT.

Kaarstad, M., Rindahl, G., Torgersen, G.-E. and Drøivoldsmo, A. 2009. Interaction and Interaction Skills in an Integrated Operations Setting, in *Proceedings of the 17th Congress of the International Ergonomics Association*. Beijing, China.

OLF (Norwegian Oil Industry Association), 2008. *Integrated Operations in New Projects*. Available at http://www.olf.no/PageFiles/14295/081016_IOP_PDF.pdf [accessed 2 May 2012].

OLF (Norwegian Oil Industry Association), 2011. *090 – OLF Recommended Guidelines for Common Model for Safe Job Analysis (SJA)*. Available at http://tinyurl.com/ccrr7jp [accessed 16 July 2012].

Rasmussen, J. 1983. Skills, rules, knowledge; signals, signs, symbols, and other distinctions in human performance models. *IEEE Transactions on Systems, Man, and Cybernetics* SMC-13 (3), 257–266.

Rindahl, G. Torgersen, G.-E., Kaarstad, M., Drøivoldsmo, A. and Broberg, H. 2009. *IO Collaboration and Interaction at Brage. Collecting the Features of Successful Collaboration that Training, Practices and Technology must*

Support in Future Integrated Operations, IO Center Report No. P4.1-003. Trondheim: Center for Integrated Operations in the Petroleum Industry.

Sarshar, S., Rindahl, G., Skjerve, A.B., Braseth, A.O. and Randem, H.O. 2010. Future Collaboration Environments for Risk Informed Decisions. An Explorative Study on Technology for Mapping of Information and Knowledge onto a Shared Surface to Improve Risk Identification, in *Proceedings of the 2010 IEEE International Conference on Systems, Man, and Cybernetics*. Istanbul: IEEE.

Skjerve, A.B., Rindahl, G., Randem, H.O. and Sarshar, S. 2009. Facilitating Adequate Prioritization of Safety Goals in Distributed Teams at the Norwegian Continental Shelf, in *Proceedings of the 17th Congress of the International Ergonomics Association*. Beijing, China.

Skjerve, A.B., Rindahl, G., Sarshar, S., Randem, H.O., Fallmyr, O., Braseth, A.O., Sand, T. and Tveiten, C. 2011. *The Integrated Operations Maintenance and Modification Planner (IO-MAP) – the First Usability Evaluation – Study and First Findings*. IO Center Report No. P4.1-004, February 2011.

Storey, N. 1996. *Safety-Critical Computer Systems*. Harlow: Pearson Education Limited.

Tufte, E. 1983. *The Visual Display of Quantitative Information*. Cheshire, CT: Graphics Press.

Tufte, E. 1990. *Envisional Information*. Cheshire, CT: Graphics Press.

Tufte, E. 1997. *Visual Explanations*. Cheshire, CT: Graphics Press.

Chapter 11
Lessons Learned and Recommendations from Section Two

Denis Besnard and Eirik Albrechtsen

Take-home Messages from Chapters 6–10

Vatn and Haugen (Chapter 6) discussed the usefulness of risk analysis, taking the Deepwater Horizon accident and the Gullfaks C incident as examples. The main operational messages are as follows:

- Risk management at the sharp-end is improved by better integration of the various categories of risk analyses and the systematic use of a risk matrix during the entire lifetime of an installation.
- Greater attention should be given to the principal factors that influence risk rather than focusing on a description of each and every causation path.
- More comprehensive scenario models should be developed, that can form a common platform for all three types of risk analysis (strategic, qualitative and operative).
- Integrated Operations (IO)-based solutions may improve the usefulness of risk analysis by providing access to onshore expert centres and more up-to-date risk analyses.

Besnard (Chapter 7) presented a human–machine interaction (HMI) assessment method for IO-based solutions to improve drilling and well operations. The main operational messages are as follows:

- There is great potential for knowledge transfer from HMI in aviation (and other domains) to oil and gas.
- The automation of HMI can lead to various pitfalls that may degrade performance in remote control and support of offshore drilling operations.
- A list of questions can be used to assess these HMI pitfalls, and the resulting assessment can be scored.

Apneseth et al. (Chapter 8) documented the use of the Resilience Analysis Grid (RAG) for Integrated Planning (IPL). The main operational messages are as follows:

- The RAG can be applied to planning in the oil and gas industry to identify the critical functions of activities and evaluate their resilience.
- During the survey, the generic dimensions of each cornerstone must be adapted to the system at hand, and can be assessed on an ordinal scale.
- The RAG provides a snapshot of the resilience profile of an organization for a given task or project.

Tveiten (Chapter 9) outlined the use of the Functional Resonance Analysis Method (FRAM) to analyse unwanted consequences related to modifications to oil and gas installations. The main operational messages are as follows:

- FRAM identifies latent failures and potential variability in the planning phase of high-risk operations.
- FRAM leads to the development of countermeasures to dampen unwanted performance variability in a system, which might lead to unwanted consequences.
- FRAM is also well-suited to the study of normal working conditions in complex operations.

Rindahl et al. (Chapter 10) discussed the IO-MAP tool, designed to contribute to safer decision-making in collaborative environments. This tool facilitates hazard identification by means of visualization using shared collaboration surfaces. The main results of their usability study were as follows:

- The tool helps workers to identify hazards more effectively.
- It also improves communication about the location and nature of activities.

Recommendations

On the basis of these chapters, the following operational recommendations can be made:

- The design and use of the methods and tools documented in this section demonstrate knowledge transfer from academia to risk managers in the oil and gas industry. This transfer is promoted by a) academics tackling industrial challenges, b) companies supporting the in-house customization of a prototype and c) risk managers being ready to add a new method or tool to their toolbox.
- The introduction of IO-based solutions demands modifications to generic risk control and assessment methods. It follows that IO must comply with basic engineering rules: generic methods must be customized and validated in settings and activities that are specific to the target company.
- The methods described in this section offer new ways to assess and visualize safety-related issues. The addition of supplementary methods to the risk management toolbox enables a broader perspective on safety issues, which strengthens support for decision-making.

Section Three
Risk Assessment of an IO Scenario
from Different Perspectives

Chapter 12
Risk Assessment in Practice: An Integrated Operations Scenario from Two Different Perspectives

Eirik Albrechtsen

This chapter describes a fictitious scenario related to oil and gas drilling operations. The scenario was created in order to show how different theoretical perspectives can be applied to risk assessment in Integrated Operations (IO). It is used in the following chapters of this section to demonstrate two different approaches to risk assessment, namely quantitative risk assessment and Resilience Engineering. In addition to outlining the basic steps and principles, the authors elaborate on the methods used, describe how risk is assessed and expressed (according to the needs of the decision-makers and/or the processes involved) and the main assumptions and limitations.

Objectives

The following two chapters present the results of these two approaches. In Chapter 13 Jørn Vatn shows how knowledge can be structured to express risks embedded in the scenario. In Chapter 14 Erik Hollnagel shows how a Resilience Engineering perspective can be applied to risk assessment. Then in Chapter 15, Eirik Albrechtsen and Denis Besnard discuss the two approaches provided by Hollnagel and Vatn and reflect on links and differences between them.

The purpose of the exercise is to demonstrate how theoretical risk assessment concepts can be applied in practice. Practitioners may find it useful to compare approaches, despite the fact that no attempt is made to recommend one approach over the other in terms of what method to choose, in what circumstances, and so on. Nevertheless, there are some important messages to keep in mind. The first is that an analysis of the risks of technological change must take into account the organization, its staff and their work. A comparison of the results of these two, very different methods applied to the same scenario, demonstrates that risk assessment at the organizational level can be achieved using more than one method. This is in itself an important point as it emphasizes that different approaches can be complementary. Although this complementarity is valuable in risk management it can suffer from various biases that influence the choice of method. These include the tendency to always favour the recommended in-house method, rather than one that may be more suited to the case at hand. Another example is to adopt a new, high-visibility and popular method rather than one more suited to the needs of the company. Finally, a company may select a particular method because it has the professional competence to implement it, rather than one that is suited to the company's safety requirements.

This chapter describes the scenario used for the exercise. The description is based on a case study developed in a workshop that included participants from the industry, regulatory authorities and researchers. This case study can also be found in Tveiten et al. (2008).

The Scenario

Background

By the year 2017 the oil and gas field in question will have been in operation since 1980 and will be in the tail production phase.

The operators, a global oil and gas company, plan to move the central control room (CCR) on an existing installation from offshore to onshore. Various investigations have suggested that remote operation would improve both the efficiency and safety of the platform. The company has already established several onshore expert centres, which support operations and will continue to support the onshore CCR. A control room, manned by a single operator during daylight hours will continue to operate on the platform. The CCR also manages a subsea installation via remote control.

The decision to move the CCR onshore was made difficult by the opposing views of the operating company and labour organizations. The latter opposed the decision on the grounds of safety. The operator and contractors also expressed concern about the move on safety grounds, particularly related to the night-time period during which the offshore CCR is unmanned.

In recent years a large number of offshore workers have retired and it has been difficult to recruit new staff. Consequently there are very few staff on the platform or in the CCR who possess significant offshore experience. Local recruitment problems are also the reason why a number of offshore engineers have been recruited from overseas. These staff naturally bring with them a different language and culture. Another concern is that it may be difficult to persuade staff at the onshore operations centre to switch to offshore jobs once they have onshore experience; at the same time there is a worry that this may result in the loss of hands-on experience in onshore centres.

The challenges identified so far include differences in culture and language, offshore/onshore shift arrangements and the increasing number of handovers. A further challenge that has a significant impact on control room operators is the constant pressure from various parties to increase production.

Overview of the organization

Following implementation of the onshore CCR, the system is expected to look as follows.

The organization

- The operator is a global company, the head office is in the United States and it has installations in various oil and gas fields worldwide. There is a national office in Norway that is responsible for operations on the Norwegian Continental Shelf.
- Lean staffing: at night production will be remotely operated.

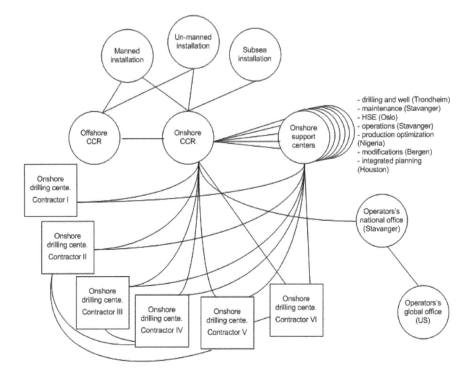

Figure 12.1   Actors involved in the operation of the onshore CCR

- Three installations: one manned, one unmanned and a third on the ocean floor.
- Figure 12.1 shows the actors involved in the operation of the CCR.

The onshore central control room

- All control room functions remain with the onshore CCR in Norway.
- The onshore CCR is fully operational.
- It is manned in three eight-hour shifts per day, seven days per week. There are three operators in each shift.

The offshore central control room

- Fully operational offshore control room; manned by one operator in 12-hour day shifts.
- Closed at night.

- Manned by four offshore process operators, one of whom is available to the onshore CCR at night if needed.
- Offshore process operators work day and night shifts and are all cross-trained.

Working hours, shift arrangements and rotation system

- Offshore: a 12-hour day shift; two weeks offshore followed by four weeks off.
- Onshore: three, eight-hour shifts. Onshore staff rotate every nine months. Staff working in the operations and planning group have standard eight-hour workdays.

Onshore support centres

- There are seven support centres located worldwide that provide expert knowledge. They have access to real-time information about the performance of the installation in question as well as information from the operator's other installations.
- The support centres assist not only the installation in question, but also all the operator's other installations, both on the Norwegian Continental Shelf and in other countries. For example, the maintenance support centre, responsible for maintenance planning (opportunity-based maintenance), also supports operations in Africa, the Caspian Sea and the Gulf of Mexico.
- The support centres are manned 08.00–16.00 local time.
- A team of experts located in Nigeria help to interpret the steadily increasing volume of reservoir data. The CCR communicates with these experts on a daily basis in collaboration rooms (via broadband communication). These experts also offer well service support during the night-time period.

Contractors

- Five contractors are currently involved in various field operations.

- Collaboration is based on the concept of integrated contractors. Collaboration technology makes it possible to integrate operators and suppliers, for example in morning meetings. Contractors share the same real-time information as the operator and have their own collaboration rooms that are used for daily communication with the operator.

Power supply

- The platform's energy is supplied from onshore. There is an emergency on-board generator in case of a power outage. Satellite communication provides backup in the event of a breakdown in ordinary lines of communication.

References

Tveiten, C.K., Lunde-Hanssen, L-S., Grøtan, T.O. and Pehrsen, M. 2008. *What is Actually Implied by Integrated Operations?* SINTEF Report No. A8481. Trondheim: SINTEF.

Chapter 13

Steps and Principles for Assessing and Expressing Major Accident Risk in an Integrated Operations Setting

Jørn Vatn

This chapter discusses basic steps and principles for assessing and expressing the risk of major accidents in the context of Integrated Operations (IO). The scenario described in Chapter 12 provides the background for the approach, which deals with two principal aspects of the situation. First, the more or less continual introduction of new work processes referred to as IO, then a closer focus on the decision of whether to move the central control room (CCR) onshore.

Risk is defined here as uncertainty regarding the occurrence and severity of undesired events. This definition is independent of system characteristics or the context for a decision. This does not mean to say that 'one size fits all', as the methods and techniques to be applied depend on system characteristics, but rather that the interpretation of risk – in terms of uncertainty – does not change. Risk management is the process of (1) identifying hazards and threats; (2) deriving corresponding undesired events; (3) assessing the risk; (4) identifying and evaluating measures to reduce risk; (5) taking decisions; and (6) implementing measures to reduce risk and monitor system performance at various levels.

It is important to recognize that risk analyses can be conducted during different operational phases and that their format and time horizon can vary depending on the decision support that is needed. Operative risk analyses are conducted within a very short time span and are usually provide decision support for a specific work operation. A safety job analysis (SJA) is a typical example of an operative risk analysis, where the aim is to identify hazards associated with a specific task, for example a well workover (see also Chapter 6 by Vatn and Haugen for a description of different types of risk analyses). On the other hand, strategic risk analyses provide strategic support. Whether to move the CCR from offshore to onshore is an example of a strategic decision, and this chapter discusses the use of strategic risk analyses to support decision-making.

Fundamentals and Risk Definition

To assess risk and to use the result as a basis for decision support it is important to have a precise interpretation. The definition of risk used in this chapter follows that of Vatn (2012), which includes important concepts such as cause and effect, uncertainty and value. While many interpretations of risk define it as a property of the system being analysed, here risk is interpreted as uncertainty. More precisely, risk is defined as 'uncertainty regarding the occurrence and the severity of unwanted events'. This conceptual definition of risk forms the basis for the risk analysis, but a second definition is required to express risk in quantitative terms. This second definition uses probabilities.

Probability p is the likelihood of the occurrence of an event e. S is the severity of the event. Note that S is a multidimensional random quantity representing various dimensions such as the safety of personnel, environmental impacts, material damage, economic aspects and so on. A probability distribution function is used to reflect the uncertainty of S. As there is more than one event to be analysed, i is used to index all relevant events and an operational definition of risk is the set of all relevant triplets, described as:

$$R = \{<e_i, p_i, S_i>\} \mid \mathcal{D}, \mathcal{U}, \mathcal{V} \tag{1}$$

where $\mathcal{D}$ represents dialogue and risk communication with stakeholders in order to gather information about values and preferences. For example, in the scenario described in Chapter

12 employees have expressed concerns about safety if the CCR is moved onshore. These concerns needs to be understood in greater detail – are they related to gross accident risk, a threat to the working environment or occupational accidents? $\mathcal{U}$ represents information, theories and assumptions that form the basis of the understanding of the risk assessor when risk is assessed, and finally $\mathcal{V}$ represents the result of the verification processes, for example, third party verification.

Risk Assessment in an Integrated Operations Context

According to our definition, risk expresses uncertainty; therefore the risk assessment should also focus on uncertainties. One category of uncertainty concerns the completeness problem, that is, the extent to which all undesired events can be identified. A study by Skjerve et al. (2008) indicated that although the introduction of IO brought new safety challenges, the nature of hazardous events did not seem to change (for example, hydrocarbon leakages, fires, explosions, blowouts and loss of structural integrity). Therefore, in the context of the definition of risk given in equation (1) the identification of all undesired events, e_i is not seen as a major challenge. It should be noted that even though the risk analysis framework provides techniques to identify undesired events, previous knowledge of such events is always a useful starting point.

Another category of uncertainty is the ability to identify the causes of an undesired event. In this case, the major challenge is that it is impossible to prescribe in detail all possible patterns that may lead to a particular undesired event. Despite the progress that has been made, the analyst can still be surprised by new causes of failure. However, we should not be surprised that there are still surprises and the possibility should be reflected both in uncertainty statements, and particularly in the documentation of the failure cause identification procedure forming part of $\mathcal{U}$.

A wide range of modelling techniques must be deployed in order to establish a realistic quantitative model that formally expresses the various elements representing uncertainty in the analysis. These range from formal probabilistic techniques to softer approaches such as Bayesian Belief Nets (BBN); see Mohaghegh et al. (2009) and Vatn (2012) for a fuller discussion.

Primary accident scenario modelling

Risk modelling begins by structuring accident scenarios related to the undesired events that form the focus of the investigation. It consists of:

1. The identification of primary hazards and related trigger events. A primary hazard is an energy-related hazard, that is, loss of control of potentially harmful energy sources. It would appear the introduction of IO will not create new hazardous event types (Skjerve et al., 2008).
2. Modelling the corresponding event scenario using formal probabilistic risk analysis (PRA) techniques that describe the system using a logical construct, for example, fault and event tree analysis. This part of the analysis aims to model the event scenario up until the point where control of energy-related sources is lost.
3. Following an accident, apply energy-related impact techniques (these model the situation following a loss of control of potentially harmful energy sources; they include gas dispersion, ignition, fire and explosion and structural integrity modelling) and recovery analysis techniques (which model the recovery process following an emergency situation) to model the situation.

Note that Steps 2 and 3 together form an entire accident scenario and the challenge is to combine the various elements of the scenario into one formal model.

Risk influencing modelling

The primary risk model is largely composed of logical structures such as fault and event trees. Roughly speaking this part of the model describes linear causes and effects structured using formal logical statements. For example, 'gas release' AND 'ignition source' results in a fire or an explosion for gas concentrations between lower and upper explosion limits. At the same time, 'softer' relations cannot be easily expressed with such formal, logical structures. An example is where the competence of an operator is assumed to influence the probability of error when performing a critical task. To overcome the problem, Risk Influencing Factors (RIFs) and Performance Shaping Factors (PSFs) are often used to structure this step. RIFs can be modelled on two levels (see for example Vinnem et al., 2012). The first level models the direct impact of a RIF on a basic event in, for example, a fault tree, while the second level (typically the management level) is indirect and influences basic events via the first level RIFs.

Change analysis

A risk analysis seldom starts from scratch. Usually, existing risk analyses form the starting point. To make best use of existing analyses, a change analysis is often carried out that identifies the main differences between the two analyses. The change analysis often needs to be conducted on two levels; the first related to the actual decision to be made, that is, whether to move the control room onshore or not. The second level concerns general changes that are not related to the actual decision – for example the fact that existing risk analyses have not taken into account the introduction of IO. The change analysis can identify issues that need to be handled according to how they differ from the risk analysis. The following lists the main categories of differences:

1. Differences that need a qualitative rework of the formal probabilistic techniques, for example, the need to introduce new safety barriers into the model.
2. Differences that relate to the RIFs that have been identified, or the need to introduce new RIFs into the model.
3. Differences that directly change the parameters of the model, for example, failure rate.
4. Differences that cannot easily be incorporated into the risk analysis, and require a broader systemic perspective, that is, a complexity analysis.

Often, the change analysis initially identifies qualitative differences, which are studied in more detail in a second step that quantifies influences impacting the risk analysis model. The following section describes a more general approach to complexity analysis. As IO is often associated with increased complexity the approach incorporates qualitative differences into the analysis whenever possible, in order to reduce the number of steps in the analysis (see for example, Chapter 4 by Grøtan for a discussion of sensemaking in complex environments).

Complexity analysis

In this phase particular attention is paid to complexity. Complexity is used in numerous scientific fields and has no unique definition. In this chapter complexity is understood as a system composed of many parts or elements where the interactions between them are not clear. This interpretation follows that of Perrow (1984). Complexity analysis consists of the following main steps:

1. identify complexity characteristics;
2. complexity impact analysis;
3. complexity structuring;
4. complexity quantification.

Identify complexity characteristics

The literature describes a wide range of system characteristics that are associated with complexity; these include the proximity of subsystems and components, interconnected subsystems, substitutions, flexibility, time constraints, common mode connections, feedback loops, slack, few signs and intractability. It should be noted that according to Perrow (1984) some of these elements represent tight coupling rather than complexity, but for the purposes of this analysis the distinction is not important. A list of keywords describing complexity characteristics can be valuable in the initial phase of the complexity analysis. IO is often thought to introduce complexity due to the large number of actors, greater use of critical task planning on short timescales, and collaboration across geographical and organizational borders. Examples of these kinds of complexity will be seen in the analysis of the sample scenario, both in terms of critical latent conditions that are introduced as a consequence of distributed team planning, and in the identification of side effects in the critical problem-solving phases.

Complexity impact analysis

The aim of the complexity impact analysis is to establish the impact of each of the identified complexity characteristics in one or more accident scenarios. Complexity characteristics are associated with various aspects of each accident scenario. This leads to the structuring step of the complexity analysis. As an example, organizing planning according to time constraints is a complexity aspect that is not only treated in general, but also associated with the planning of a well operation. Then, in the structuring phase, the qualitative impact of the complexity characteristic is assessed in relation to the risk model. The following structuring techniques provide a starting point:

1. Extend or modify the list of RIFs acting on single parameters in the risk model. For example, in a gas leakage scenario, interactions between greater numbers of actors may increase the probability of failure in the work permit system.
2. Extend or modify the list of common cause failures (β). One way to treat time constraints would be to identify basic events in a sequence where any time constraint will affect them all, and exploit this in terms of the β-factor.
3. Introduce new, or modify existing system dynamics (SD) modelling. SD is generally considered to be efficient to model time constraints and feedback loops. This is an alternative to more implicit modelling using BBN or β-factors.
4. Add new accident scenarios.[1]
5. Extend existing accident scenarios. This particularly relate to crisis situations, where an understanding of complexity might indicate other aspects relevant to recovery.

---

1   It is important to note the distinction between types of hazardous event and accident scenarios. Although the list of hazardous event types is not expected to change following the introduction of IO, there may be more scenarios leading to these hazardous events, especially given the argument that the introduction of IO creates greater complexity.

Complexity structuring

The structuring phase of the complexity analysis is primarily a qualitative analysis where knowledge and understanding are mapped into the qualitative part of the risk analysis model.

Complexity quantification

In the quantification step a 'strength' is assigned to the various qualitative complexity characteristics. There are a variety of quantification approaches. Although the impact of complexity (as such) on risk is difficult to assess, this step helps to make the various elements more explicit. The literature identifies a range of methods that link complexity characteristics to, for example, the common mode failure issue (β-factor).

Manifest and latent complexity

There has been much discussion of complexity in the context of risk assessment. Grøtan (Chapter 4) suggests a distinction between manifest and latent complexity. Manifest complexity, as the name suggests, is to a greater or lesser extent clear. It is theoretically possible to identify this type of complexity both in advance and in retrospect. Many accident and incident investigations highlight manifest complexity. For example the investigation following the Snorre A gas blowout clearly states that the complexity of the well and the lack of well integrity were well-known issues (PSA, 2005). The audit of the loss of well control at Gullfaks C also highlighted manifest complexity (PSA, 2010). In Chapter 4, Grøtan argues that in such cases it is reasonable to talk about manifest complexity as if it is possible to expect the unexpected. On the other hand, Grøtan views latent complexity as a kind of pathogenic property of analysable systems (linear cause and effect) that presents itself very abruptly and violently. Latent complexity thus represents a situation where the unexpected cannot be expected. Latent complexity must be addressed in the risk assessment; the analysis must consider whether there are reasons to expect surprises even though there is no sign of them. A more detailed discussion of manifest and latent complexity can be found in Vatn (2012), which focuses on the Gullfaks C incident mentioned earlier.

Application of the Approach

This section discusses the application of the risk analysis approach to the scenario described in Chapter 12. Two principal issues are discussed; the first is the more or less continuous introduction of new IO work processes. The second focuses on the specific decision of whether to move the CCR from offshore to onshore. Although this decision essentially represents a go/no-go situation, the analysis also reveals other issues that require lower-level decisions, for example, actions that support opportunities and mitigate threats related to the move. Various decisions are taken by different parts of the organization. For example, the decision to move the CCR or not is typically taken by the board who do not usually look at detailed safety management issues. They simply want to know if the move can be achieved without compromising safety. On the other hand, the project team responsible for carrying out the move need to have a far more detailed understanding of the consequences of specific decisions, such as the training required to maintain safety functions.

We have defined risk as $R = \{<e_i, p_i, S_i>\} \mid \mathcal{D}, \mathcal{U}, \mathcal{V}$ which takes into account the need for decision support and the concerns of stakeholders. In the case study, there are two decisions to be made. The first is whether the risk of moving the CCR onshore is acceptable, and the second concerns the potential implementation of measures to reduce any negative impact or support the positive impact of RIFs or conditions.

It is important to note the labour organizations have expressed concern about the potential safety impact of moving the CCR onshore. Employees are an important stakeholder in the dialogue processes ($\mathcal{D}$) and it is essential to have their agreement on the general focus. However, experience has shown that the dialogue processes often identifies specific threats and hazards of concern to stakeholders. A general safety concern often translates into more specific concerns, for example, loss of skills as a result of the move to control room operator. Making general safety concerns explicit makes it much easier to hold frank discussions about specific concerns, and to identify how to handle them in the risk analysis.

The result of the risk analysis is a list of identified undesired events, together with corresponding uncertainties described in terms of likelihood of occurrence and severity ($<e_i, p_i, S_i>$) both for the current situation and for the post-move situation. Some example of events that the analysis may focus on include: lack of control of the well stream, gas leakage, fire, falling objects, and overpressure in the processing unit. For some of these events the situation may become more favourable and for others, less favourable. An assessment of the total impact of risks must aggregate the expected values, that is, the total probabilities of events with a given severity. This should always be carried out on the basis of any issues discovered during dialogue processes ($\mathcal{D}$). It should also be noted that in this approach, risk is always conditional on the underlying assumptions, understanding, knowledge base and available empirical data ($\mathcal{U}$). Therefore it is important to document $\mathcal{U}$ as thoroughly as possible and in a format that can be understood by non-experts in risk analysis. A sensitivity analysis of critical assumptions should also form part of the documentation. For example, it might be reasonable to assume that the physical conditions in the well will not change; hence failure rates for downstream components can also be assumed not to change. When this assumption is documented, it can be questioned. If this reveals a doubt, a further analysis might be required to identify any changes related to the well stream.

Primary accident scenario modelling

In order to provide adequate support for decision-making, a wide range of scenarios would normally be assessed; this chapter will only focus on one. Following the disaster in the Gulf of Mexico (OSC, 2011) and the critical event on the Snorre A platform (PSA, 2005) the analysis will develop a scenario involving loss of control of the well stream.

Norwegian legislation requires there to be two independent barriers against loss of well stream control and that the status of these two barriers is continuously monitored. As it is not possible to model these requirements explicitly in a strategic risk analysis, the strategic risk analysis must be linked to an operative risk analysis. The main objective of the operative risk analysis is to document the performance of the two barriers and any threats that may compromise barrier integrity throughout the entire operation. The investigation that followed the Snorre A event showed that the integrity of the well had already been compromised prior to the critical well operation. Although this was known, appropriate measures were not implemented. Unfortunately, this is not a unique situation and a full analysis would address the scenario where there are known threats to one or more of the barriers, but where it is uncertain whether the threats will occur and, if they do, whether there are adequate measures to handle them.

Figure 13.1 shows the event tree developed to describe the well operation scenario. An event tree shows the possible course of events following an initiating event. Each box in the diagram contains a statement regarding an event or a situation. If the statement is 'true' (☺) the diagram is traversed vertically, while if it 'false' (☹) the diagram is traversed horizontally. The diagram indicates where the initiating event and two other events are analysed in more detail by a fault tree analysis (FTA). It should be noted that the event tree shown in Figure 13.1 is generic and does not focus on a specific latent condition. Ideally the event tree would be developed for a wide range of latent conditions. In practice, this is not feasible and only a limited number of specific scenarios are developed. The risk analysis can therefore never provide comprehensive coverage of every aspect[2] and the extent of coverage depends on the need for decision support in a given context.

---

2   The fact that a risk analysis is never complete may be seen as a weakness of the approach and is used as an argument to promote other approaches. However, there is no reason to believe that other approaches provide additional insight. As by definition, they do not explicitly treat uncertainty, uncertainty arguments are not fully taken into account. Although other approaches can be a useful part of the risk identification process, they should be integrated into the method and not seen as alternatives. It is also important to emphasize that the risk analysis approach is iterative, the various factors and issues of concern can be ranked and analysts can therefore focus on the most important issues.

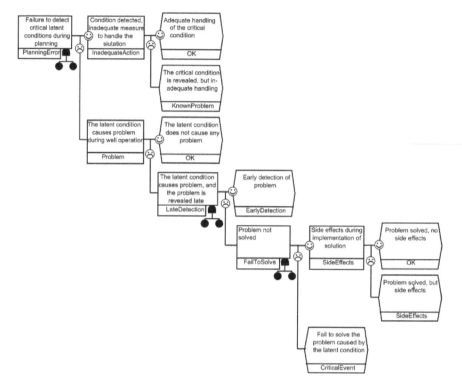

Figure 13.1   Primary event tree – critical latent conditions prior to well operation

The bottom row of the fault tree shown in Figure 13.2 shows three basic events that will all result in a failure to identify the problem. The investigation following the Snorre A event revealed that operative risk analyses was first postponed and then cancelled. The failure to properly identify hazards also applies to other accidents. For example, the railway accident at Sjursøya (AIBN, 2010) highlighted the fact that a risk analysis was begun several years prior to the accident but never completed. Other accident investigations have shown that although critical conditions or problems were known at an early stage, they were not formally documented (in a risk register or similar instruments for sharing knowledge). It should be noted that event and fault tree analyses are formal PRA techniques and the qualitative part of the analysis is very often independent of the specific situation being studied. The individual elements that make up events are often generic and independent of the actual situation or decision

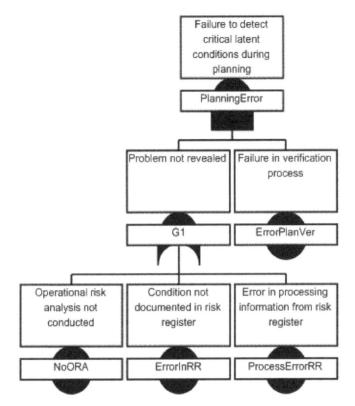

Figure 13.2   Fault tree for the event 'Planning Error'

to be made. Issues that are specific to the particular scenario are included in a later stage, for example, risk influence modelling.

The fault tree shown in Figure 13.3 models the period from the point when the latent condition becomes a problem to the moment it becomes apparent. Note that FailToAnticipate is included as a basic event, but at this stage of the analysis it is not linked to influencing factors, for example, lack of situational awareness or geographically distributed actors. Ideally a stochastic variable should be used to model this temporal aspect and a dynamic event tree analysis should be carried out. However, to simplify the example the condition is said to be detected either 'early' or 'late'. Note that while the fault tree shown in Figure 13.2 represents the planning phase (prior to actual well operation), the fault tree shown in Figure 13.3 represents an early stage of the operative phase. Finally, the fault tree shown in Figure 13.4 represents the problem-solving phase. Note the AND gate (G2)

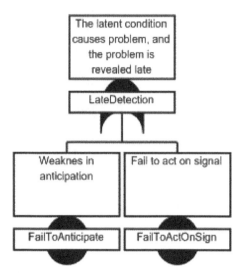

Figure 13.3   **Fault tree for the event 'Late Detection of Critical Situation'**

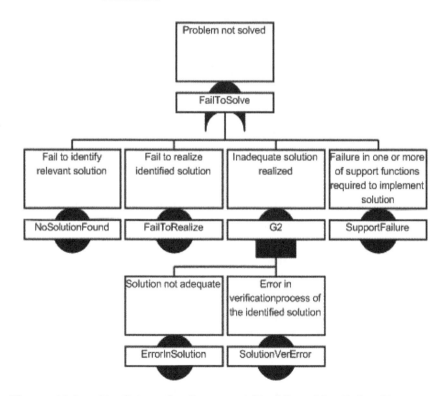

Figure 13.4   **Fault tree for the event 'Problem Not Solved'**

where SolutionVerError eliminates proposed solutions that are inadequate. It should also be noted that problem-solving is an iterative process and the fault tree representation may be replaced by another model (for example, a system dynamic model) if required.

Modelling risk influencing factors

The literature provides a wide range of RIFs that can be included in the model. Although many studies have focused on scenarios similar to the event tree shown in Figure 13.1 (for example, Vinnem et al., 2012) none of these studies have focused on IO. Therefore, the following is not a comprehensive list of RIFs, but is rather a list of RIFs that are of particular interest in relation to IO:

1. Coordinating Capabilities Planning (CCP) are coordination capabilities in the planning phase. It describes the ability to extract and process information from many different actors upstream of critical operations. Formal instruments include risk registers, coordination meetings and documented work plans approved by the relevant players (RIF_{CCP}).
2. Coordinating Capabilities Operation (CCO) are coordination capabilities in the operational phase. It describes the ability to quickly obtain accurate knowledge from available experts and combine knowledge and information to provide appropriate decision support (RIF_{CCO}).
3. Common Shared Awareness (CSA) is the ability to quickly achieve an accurate shared awareness of a problem. It involves anticipation, and capabilities/training in creating mental models of the situation in distributed teams (RIF_{CSA}).
4. Problem Solving and Improvisation (PSI) describes problem-solving capabilities such as the capacity for creative search and improvisation (RIF_{PSI}).
5. Local Knowledge (LK) is the knowledge of the operational team. Historically the offshore CCR served as a meeting place where knowledge was shared by the team through various informal processes – it provided the metaphorical 'camp fire' (RIF_{LK}).

6. External Knowledge (EK) describes access to external knowledge. In both a critical situation and in the planning phase of critical operations, access to expertise is recognized as a critical factor (RIF_{EK}).

It should be noted that the various RIFs will impact different elements in formal probabilistic models (event tree and fault trees). For example, the main impact of RIF_{CCP} will be on the basic events of the fault tree shown in Figure 13.2, while the main impact of RIF_{CSA} will be on the basic events of the fault tree shown in Figure 13.3, and the main impact of RIF_{CCO} and RIF_{PSI} will be on the basic events shown in the fault tree in Figure 13.4. It should also be noted that a RIF that is positive in one context may be negative in another. For example, although RIF_{CCO} may be positive with respect to the fault tree shown in Figure 13.4, it may be negative in the context of the fault tree shown in Figure 13.2. The investigation into the Snorre A event revealed that the crew had very good problem-solving capacities and local knowledge that was crucial to salvaging the situation. However, the investigation also highlighted that the confidence the Snorre A team had in their problem-solving abilities was problematic. Overconfidence in their ability to improvise compromised upstream activities that would have ensured safe work processes without having to improvise. Although generally, improvisation and creativity are believed to be positive capabilities in critical situations, they are not always appreciated in the upstream planning phase.

Change analysis

IO generally results in changes to work processes. One challenge is the increased number of actors located in various countries around the world, and this situation is clearly more complex than traditional modes of operation. It is therefore reasonable to argue that the RIF_{CCP} (planning coordination capabilities) will be unfavourably affected by the introduction of IO. It should also be noted that planning and coordination activities prior to well operation not only affect the integrity of well barriers, but also the performance of support functions.

The following hypothetical situation demonstrates the point. One of the two mud pumps at an installation is not working properly and is awaiting repair. This activity is managed by the maintenance contractor in Norway who faces the challenge of communicating the problem to the well optimization team located in Nigeria. One way to model the relation would be to add a virtual common mode basic event to the fault trees shown in Figure 13.2 and Figure 13.4 respectively and have this virtual event influenced by RIF_{CCP}. In this case, the change analysis has not only revealed a potential weakening of a RIF, it also provides an argument for a qualitative update to the formal logical risk model by introducing a common mode event to represent systemic aspects. In the same way, other aspects that may change as a result of the introduction of IO can be introduced. Among these, an important issue to take into account is CSA. However, it is not possible to determine whether the RIF_{CSA} will change in principle.

The next part of the change analysis deals with the impact of moving the CCR from offshore to onshore. The Snorre A event emphasized that local knowledge was a factor in the successful resolution of the incident. A clear area of concern is therefore any changes to RIF_{LK} (local knowledge). As pointed out earlier, the offshore CCR served as a meeting place where experience and knowledge was shared between crews and shifts. The move to an onshore CCR is likely to weaken RIF_{LK} unless countermeasures can be implemented to maintain the 'camp fire' function. Another factor that may undermine RIF_{LK} is the specific challenge of the loss of key personnel that may come about as a result of the reorganization.

Another area of concern is RIF_{CCO} (operational coordination capabilities). Although in the scenario, a backup CCR will remain offshore this CCR cannot be expected to have the same operational coordination capacity as before. In a critical situation there will therefore be less face-to-face contact between CCR staff and operational teams. The investigation report of the Snorre A event emphasized how operational teams were able to reconfigure the air inlet to the mud pumps when gas was discovered. This raises the question of the extent to which it is possible to supervise events offshore from an onshore CCR.

The move to an onshore CCR may also improve RIF_{CCO}. The onshore CCR is expected to have a greater capacity to retrieve knowledge from expert centres throughout the world, greater database search capabilities (for example, data mining and case-based reasoning) and improved pattern recognition of documented problems, and so on.

Modelling raises various issues (see Vinnem et al., 2012 for a more detailed discussion). First of all, it is important to distinguish between the importance of a RIF and its value. Importance is essentially a measure of the impact an un/favourable RIF has on event and fault tree parameters. One way to model it is to introduce sensitivity factors at the parameter level. For example a (symmetric) sensitivity factor of five means that a failure probability is assessed to be five times higher than the average value if the related RIFs reach the worst possible state, and five times lower if the RIFs reach their best possible state. Factor regression-based techniques may be applied to assess the sensitivity factor. Regression-based techniques are simple to apply in controlled experimental designs, but prove more challenging in real-life scenarios. It is however possible, to some extent, to obtain ranges of sensitivity factors, for example by correlating information related to gas leakages and RIF scores recorded in audits. Another approach is to review accident and incident investigation reports and try to link the importance of one or more RIFs to events described in the report. For example, if time pressure is frequently mentioned in investigations related to a particular incident, it is reasonable to assign the corresponding RIF a high sensitivity factor. Two issues must be taken into account when assessing the importance of a RIF: the current situation (prior to any decision being taken) and the situation following the change. In many cases it is difficult to assess change in the RIF; therefore it is important make any assumptions ($\mathcal{U}$) as explicit as possible.

Complexity analysis

Vatn (2012) describes a comprehensive complexity analysis in the context of opportunity-based maintenance in an IO setting. Therefore, the discussion here is limited to the need to extend existing accident scenarios.

The event tree shown in Figure 13.1 shows the positive exit (☺) from the FailToSolve event and is described schematically by a new SideEffects event. The number of actors involved in the operation (drilling, maintenance, production optimization, modification, and so on) means these side effects must be analysed more comprehensively. For example, maintenance work may be carried out at the same time as operations. However, should there be problems with operations, it may be critical to be able to abandon maintenance activities such as welding. An example of this was seen in the Snorre A event, where the air inlet for critical components required for the operation of mud pumps (needed to maintain the 'mud barrier') was located within the stream of gas escaping from the well. Luckily, in this case, the problem was identified.

The issue is not that the move to an onshore CCR would have changed the probability of the side effect, but rather the fact that the large number of actors makes it generally difficult to identify such relationships in a stressful situation. The challenge related to the air inlet was easy to identify after the event (Kurtz and Snowden, 2003), but prior to an incident the analysis can only be carried out on a generic level, for example the identification of side effects related to threats to barrier integrity, support functions, the existence of hydrocarbons, ignition problems, falling objects (on critical pipes), platform stability and so on.

One way to handle these issues would be to develop the SideEffects event into a fault tree where various generic side effects are combined with an OR gate. Each side effect could then be combined with a verification event linked by AND gates to represent barriers. Note that the performance of these barriers (for example, anticipation capabilities during the development of the critical situation) would most likely not be independent of the SolutionVerError shown in Figure 13.4. Such systemic effects would be modelled by a common mode event, where the β-factor is modelled as a function of various RIFs, e.g., RIF_{CSA}, RIF_{LK} and RIF_{CCO}.

The above only discusses some aspects of the complexity analysis. In real life, the analysis would be extended to include more events in the formal, logical part of the risk model. However, the analysis should nevertheless be limited to those parts of the risk model considered to be most affected by the decision to be taken.

Discussion

The risk analysis must be relevant to the decisions to be taken. In this case study, the main decision to be taken was whether to move the CCR onshore. Subsequent decisions relate to risk reduction measures to be implemented. The following discussion focuses on the decision to move the CCR onshore.

As only a part of the risk scenario is considered here, the calculation of the risk $R = \{<e_i, p_i, S_i>\} \mid \mathcal{D}, \mathcal{U}, \mathcal{V}$ is straightforward. Let e_i represent the existence of a latent condition during planning. Furthermore, let p_M and p_0 respectively be the estimated probability of this event if the CCR is moved or not. To simplify the example, only the worst-case outcome is considered. Therefore, let $p_{S,M}$ and $p_{S,0}$ respectively be the estimated probability of the most severe outcome assuming the event e_i occurs if the CCR is moved or not. Next, assume that under the assumptions made in the analysis ($\mathcal{U}$) moving the CCR onshore gives $p_M < p_0 \mid \mathcal{U}$ and $p_{S,M} < p_{S,0} \mid \mathcal{U}$.

The presentation of the results should include both the numerical values representing probabilities and the main arguments leading to these conclusions. For example, the need for stronger links to onshore support centres both in the planning phase and in the problem-solving phase, where fast access to support centres and improved training in utilizing their expertise is considered crucial. The results of the overall analysis should also list the principal findings of the sensitivity analysis. For example, in this study the assumption that the move to an onshore CCR would reduce the contribution of local knowledge (as there are fewer arenas for sharing experience) has been assessed. Consequently, the weighting given to RIF_{LK} in the sensitivity analysis was increased and the result of the analysis showed that the two solutions were in fact almost equal with respect to risk. Similarly, the negative weighting of RIF_{CSA} was increased in the case of the move to an onshore CCR, which when taken together with the increased weight of RIF_{LK} suggested a higher risk associated with the move to an onshore CCR, e.g., $p_M \gg p_0 \mid \mathcal{U}^1$ and $p_{S,M} > p_{S,0} \mid \mathcal{U}^1$. Therefore, from a risk point of view and using assumptions ($\mathcal{U}$), the CCR should be moved onshore. Third party verification ($\mathcal{V}$) also confirms that ($\mathcal{U}$) seems reasonable.

However, other results (for example, from the sensitivity analysis, $\mathcal{U}^1$) may influence the decision. For example, when results are presented to stakeholders such as unions, assumptions related to the dialogue process ($\mathcal{D}$) are emphasized. In this situation, although significant weight was given to the offshore experience of onshore CCR personnel, the risk assessment seems to argue in favour of moving the CCR onshore.

References

AIBN (Accident Investigation Board Norway). 2010. *Foreløpig rapport med umiddelbar sikkerhetstilråding jernbaneulykke Alnabru – Sjursøya den 24 March 2010* [Preliminary Report with Initial Safety Recommendations Following the Alnabru – Sjursøya Rail Accident 24 March 2010]. Available at http://tinyurl.com/bv5lhcw [accessed: 13 July 2010].

Kurtz, C.F. and Snowden, D.J. 2003. The new dynamics of strategy: Sense-making in a complex and complicated world. *IBM Systems Journal*, 42 (3), 462–483.

Mohaghegh, Z., Kazemi, R. and Mosleh, A. 2009. Incorporating organizational factors into Probabilistic Risk Assessment (PRA) of complex socio-technical systems: A hybrid technique formalization. *Reliability Engineering and System Safety*, 94, 1000–1018.

OSC (Oil Spill Commission). 2011. *Deep Water – The Gulf Oil Disaster and the Future of Offshore Drilling, Report to the President*. United States: National Commission on the BP Deepwater Horizon Oil Spill and Offshore Drilling. Available at http://www.oilspillcommission.gov/final-report [accessed 13 July 2012].

Perrow, C. 1984. *Normal Accidents: Living with High-risk Technologies*. New York: Basic Books.

PSA (Petroleum Safety Authority). 2005. *Investigation of Gas Blowout on Snorre A, Well 34/7-P31A, 28 November 2004*. Petroleum Safety Authority Norway. Available at http://tinyurl.com/7ywjxnf [accessed: 13 July 2012].

PSA (Petroleum Safety Authority). 2010. *Audit of Statoil's Planning for Well 34/10-C-06A A*. Petroleum Safety Authority Norway. Available at http://tinyurl.com/746h79f [accessed: 13 July 2012].

Skjerve, A.B., Albrechtsen, E. and Tveiten, C.K. 2008. *Defined Situations of Hazard and Accident Related to Integrated Operations on the Norwegian Continental Shelf.* SINTEF Report A9123. Available at http://tinyurl. com/7hg825a [accessed: 13 July 2012].

Vatn, J. 2012. Can We Understand Complex Systems in Terms of Risk Analysis? *Proceedings of the Institution of Mechanical Engineers, Part O: Journal of Risk and Reliability,* June 2012, 226 (3), 346–358.

Vinnem, J.E., Bye, R., Gran, B.A., Kongsvik, T., Nyheim, O.M., Okstad, E.H., Seljelid, J. and Vatn, J. 2012. Risk modelling of maintenance work on major process equipment on offshore petroleum installations. *Journal of Loss Prevention in the Process Industries,* 25 (2), 274–292.

Chapter 14

A Resilience Engineering Approach to Assess Major Accident Risks

Erik Hollnagel

This chapter describes how the principles of Resilience Engineering can be used to make a risk assessment of an Integrated Operations (IO) scenario. It refers to the case study provided in Chapter 12.

Resilience Engineering and Risk Assessment

Since 2004 Resilience Engineering has been developed and promoted as a complement to established safety approaches. These methods are widely and commonly used for accident investigation and risk assessment across all major industries. As many of the methods were originally developed in the 1960s and 1970s, it is hardly surprising that they reflect the characteristics of processes and industries as they existed then. Industrial processes and practices have, however, changed significantly over the last 40–50 years, not least because of the extensive – and intensive – use of Information Technology (IT) in every aspect of operations, management and communication. It can therefore not be taken for granted that methods developed for pre-IT industrial processes also will be adequate in today's complex IT-based industries and business environments. Quite apart from the rampant growth in the complexity of technology itself, today's systems are socio-technical, which means that descriptions based on technology alone are insufficient. The difference between established safety approaches and Resilience Engineering is perhaps most easily demonstrated by contrasting their respective definitions of safety.

Safety as the absence of adverse outcomes

Established approaches usually define safety in terms of the absence of adverse outcomes (accidents, incidents and so on), that is, of things that go wrong. The understanding is that a higher level of safety corresponds to a smaller number of adverse outcomes. This interpretation of what safety means is common across application domains, including offshore. For instance, the United States Agency for Healthcare Research and Quality (AHRQ, 2007) defines safety as 'freedom from accidental injury', or 'avoiding injuries or harm to patients from care that is intended to help them.' The International Civil Aviation Organization (ICAO, 2009) defines safety as 'the state in which the risk of harm to persons or of property damage is reduced to, and maintained at or below, an acceptable level through a continuing process of hazard identification and risk management'. A generic definition could be that the goal of safety is to reduce the number of adverse events, and that safety accordingly represents the organization's ability to manage the risks that are part of its operations under specific conditions and in a specific environment.

Safety as the ability to succeed

The position taken by Resilience Engineering is that safety represents the ability to succeed under varying conditions. More formally, resilience is defined as, 'the intrinsic ability of a system to adjust its functioning prior to, during, or following changes and disturbances, so that it can sustain required operations under both expected and unexpected conditions' (Hollnagel, 2011a). Safety should therefore be measured as the presence of this ability, rather than its absence. This means that safety cannot be measured in terms of the number of adverse events, of things gone wrong. Instead it is necessary to measure how effective and productive the organization is, for instance in terms of the number of things that succeed, or go right.

Safety and risk management

This difference between the established safety approaches and Resilience Engineering has consequences for the role played by risk assessment. When industrial safety is defined as 'the ability to manage the risks inherent to operations or related to the environment' (Jaubert, 2006), the identification and assessment of risks obviously becomes an important part of, and even a pre-requisite for, safety. Although the notion of risk is far from simple (for example, Aven, 2007, 2009, Okrent and Pidgeon, 1998, Klinke and Renn, 2002, Rosness, 2009, Slovic, 2001), in practice it is firmly associated with the occurrence of unwanted outcomes. Knowing what the risks are is therefore essential for safety management; both to prevent unwanted outcomes occurring and to protect against their consequences.

In Resilience Engineering, the primary goal is not to manage the possible risks, but rather to establish and manage the organization's ability to succeed under varying conditions. The emphasis is on being safe, rather than on not being unsafe. The focus is therefore on unexpected events in general, which comprises opportunities as well as risks. Indeed, it is the occurrence of the unexpected per se that is a risk to the organization, since by definition, in such cases the organization is either not able, or not ready, to respond to what is happening. Not being able to respond means that an organization does not know what the response should be, usually because the situation is new or unknown. Not being ready to respond means an organization is unable to deliver a prepared response at the right time and/or with the right intensity and duration.

Risk assessment does not play the same role for Resilience Engineering as it does for established safety approaches. Indeed, the purpose of Resilience Engineering is to manage the resilience of an organization, rather than to identify and evaluate possible risks. This position, however, needs to be qualified in relation to the concepts of direct and indirect risks.

Direct and indirect risks

The difference between safety management and Resilience Engineering can be illustrated by considering two types of

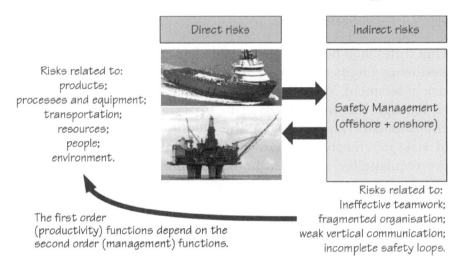

Figure 14.1   Direct (primary) and indirect (secondary) risks

risks, namely direct (primary) and indirect (secondary) risks, as illustrated by Figure 14.1. Direct risks are associated with what can go wrong in the industrial processes themselves; for instance risks related to the products that are used or made, risks related to the processes and equipment used in operations, or transportation risks related to operations. To that may be added the risks of not having the resources needed to sustain production (energy, materials, information, competent personnel), the risks that human or organizational abilities may be insufficient, the risks emanating from the physical environment (weather, tsunamis, earthquakes), and so on. The purpose of the safety management system is to identify and manage these direct risks.

A safety management system is, however, a process itself and it is therefore also necessary to consider the risks associated with it. These are the indirect risks. The most important of these is the risk that the safety management system is unable to function as required, as when this happens it becomes impossible to manage the direct risks. Resilience Engineering highlights these indirect risks, as combinations of conditions that would render the safety management system ineffective. In this sense, Resilience Engineering provides the tools and techniques needed to carry out a risk analysis of the functioning of the safety management system. Although in many cases these methods differ from those used to deal with the direct risks, the ability of Resilience Engineering to identify the indirect risks *a fortiori* means that it is also possible to identify the direct risks. This, however, is not the primary objective of Resilience Engineering, and will therefore not be considered further here.

To summarize, risk in the classical safety approaches refers to the possible failure of something – a component, a part of the system, a function or a service. The purpose of a safety assessment is to identify the risks, and try to reduce or eliminate them as far as possible. In Resilience Engineering, risk also refers to the inability of a system, specifically a safety management system, to provide the required functions. The purpose of a safety assessment is to identify such situations, or conditions, and try to find ways to ensure that the system has the ability to function in an acceptable manner in both expected and unexpected conditions.

The Four Aspects of Resilience

In order to identify the indirect risks it is necessary to provide an operational definition of what resilience entails. The essential characteristic of a resilient system is its ability to adjust its functioning so that it can succeed in everyday, unusual and difficult situations alike. This requires the four essential system abilities shown in Figure 14.2 (Hollnagel, 2009):

- Knowing what to do, that is, how to respond to regular and irregular disruptions and disturbances either by implementing a prepared set of responses or by adjusting normal functioning. This is the ability to address the *actual*.
- Knowing what to look for, that is, how to monitor what is, or can become a threat in the near term. The monitoring must cover both what happens in the environment and what happens in the system itself, that is, its own performance. This is the ability to address the *critical*.
- Knowing what has happened, that is, how to learn from experience, in particular how to learn the right lessons from the right experience – successes as well as failures. This is the ability to address the *factual*.

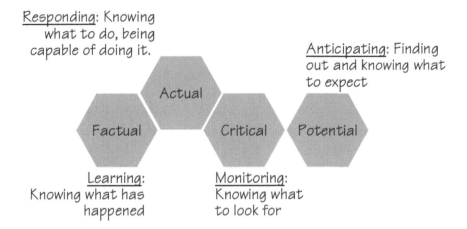

Figure 14.2   The four essential abilities of a resilient system

- Knowing what to expect, that is, how to anticipate developments, threats and opportunities further into the future. Examples include potential changes, disruptions, pressures and their consequences. This is the ability to address the *potential*.

Traditional risk assessment is, of course, a formalized way to anticipate future events, but it is confined to a subset of what could possibly happen.

Resilience Engineering makes clear that all four abilities must be present in order for an organization to be resilient. The proper balance between them depends on the application domain and the characteristics of the situation, and therefore cannot be determined a priori. Resilience Engineering also makes clear that these abilities cannot be considered independently or individually, but that they are functionally dependent on each other. Anticipation, for instance, is not possible without drawing upon past experience. Similarly, responding is not possible without predicting what the consequences of a response may be, or without monitoring how the current situation develops. Learning is required for many reasons, one being that learning reduces the probability of repeating actions that are either ineffective, or which have a negative impact on the overall system.

If the resilience of a system is defined by its ability to respond to the actual, to monitor the critical, to anticipate the potential and to learn from the factual, the obvious question for a risk assessment is 'how can these abilities be jeopardized?' The rest of this chapter will suggest how that can be done.

The essence of the method

The first step of the method is to define more precisely what the four abilities entail for a given domain and operation. This concerns both the details of the four abilities and their relative importance (Apneseth, 2010). As far as the latter is concerned, this may be determined by the nature of an organization's business. For an organization that primarily depends on exploration, be it of oil deposits or market opportunities, it is clearly important to be able to anticipate and to learn. For an organization that primarily depends on production, it may be more important to be able to monitor the situation and to respond to even minor disturbances.

As far as the details of the four abilities are concerned, they can to some extent be derived from a general description of the abilities themselves, although it is clearly necessary to adjust this to fit the specific organization that is the focus of the analysis. As a starting point, each ability can be analysed using a functional decomposition, such as a goals-means analysis or a functional resonance analysis. This will produce a list of the functions (or sub-abilities) that are necessary and sufficient to provide the ability in question. For instance, for the ability to respond, it is necessary to have defined in advance a number of conditions for which a response must be prepared. In this case, an incomplete or incorrect set of conditions may constitute a risk.

In the following, each ability has been specified by listing a number of more detailed issues, see table 14.1. For each detailed issue an associated risk is mentioned. In order to keep the present discussion manageable, the number of detailed issues has more or less arbitrarily been set to five. The purpose of this is to illustrate how a functional decomposition can be used as a basis

Table 14.1   Illustration of the relation between resilience abilities and risk

Ability	Examples of More Detailed Issues	Associated Risk or Weaknesses
Ability to respond	Is there a set of conditions (events) for which responses have been prepared?	Responses may not have been systematically prepared.
	Is the set of conditions (events) updated regularly?	Prepared responses may be out of date.
	Are the prepared responses appropriate and accurate?	Responses may be inappropriate or incorrect.
	Are there clear criteria for when a response begins and ends?	Responses may not be given at the right time or for the right duration.
	Is the response readiness ensured and maintained?	System may not be ready to respond when needed.
Ability to monitor	Has a set of (key) performance indicators been defined?	Monitoring is not systematic, important developments may be missed.
	Are the indicators chosen because they are convenient rather than meaningful?	Monitoring may be looking at the wrong things.
	Does the set include leading indicators?	Monitoring is mainly reactive; important 'signals' may be missed.
	Is monitoring planned and systematic?	Quality of monitoring may be uneven.
	Have the indicators been validated?	Monitoring may be based on wrong data and results may be untrustworthy.
Ability to learn	Is learning based only on failures (accidents, incidents)?	Organization is locked into a reactive mode of working.
	Is learning shared with other organizations?	Possibility of repeating the mistakes of others.
	Is there adequate feedback from (event) reporting?	Learning may not be fully effective. Data sources may dry up.
	Are the effects of learning verified?	The wrong lessons may be learned.
	Are the effects of learning maintained?	Lessons learned may be forgotten.
Ability to anticipate	Is anticipation a recognized need of the organization?	The organization is limited to being reactive.
	Does the organization look beyond the immediate future?	Planning may be constrained and insufficiently flexible.
	Is the organization willing to take chances (accept uncertainty)?	Organization may be unable to meet sudden changes.
	Is there a clear, strategic vision of the future?	Various groups/factions may be in conflict and resources may be wasted.
	Are the pros and cons of (organizational) changes systematically considered?	Unanticipated consequences may limit the intended effects of changes.

for thinking about risk. The second column (issues) describes possible deficiencies in an ability; the third column describes the possible consequences should the condition in column two not be fulfilled. However, this set of associated risks or weaknesses does not represent a systematic or complete risk analysis of the four abilities.

A second step of the method is to perform a systematic risk assessment of the details of the four abilities. The purpose of this is to illustrate how each of the abilities may fail, rather than to calculate failure probabilities. In accordance with the principles of Resilience Engineering, the analysis should serve to identify situations and conditions where the resilience of the organization is jeopardized. This requires a consideration not only of each of the four abilities – and sub-abilities – but also of how they are coupled, that is, how they depend on each other. The ability to respond, for instance, depends on the ability to monitor, in the sense that the detection of an impending critical situation can be used to prepare or strengthen a response. The second step can thus be seen as leading to a synthesis of the situations and conditions where the organization may be unable to function effectively. In this way the focus changes from the occurrence – and probability – of the risks, to the possible consequences of the risks in terms of how they affect the organization's performance (and resilience) potential.

A third step is to propose ways of preventing such situations from occurring. This can be done by proposing meaningful performance indicators based on the results of the analysis, and then using these to identify situations where the organization's potential may be in jeopardy. In practice this amounts to enhancing the organization's ability to monitor, but with the focus on its own condition rather than the environmental conditions. It is thus fully consistent with the principle that resilience is something that a system *does*, and not something that it *has*. Preparation for critical situations must, of course, include a plan – or a strategy – for what to do in such situations. In practice, the monitoring of performance indicators constitutes an articulated basis for the organization's management of its own performance.

Characterization of the Case Study in Terms of the Four Aspects

The case study for this chapter is described in Chapter 12. Since it is provided separately, it will not be repeated *in extenso* in this chapter.

The scenario does not refer to any specific operational situation, but to the overall organization of functions when the central control room (CCR) has been moved from offshore to onshore. The description is relatively brief and does not contain many details. The illustration of a resilience-based analysis is consequently made at the same general level, without making any supplementary assumptions about how the organization may function in detail. The analysis presented here serves primarily to illustrate the consequences of applying Resilience Engineering principles in a risk assessment. A more extensive analysis will clearly require more detailed operational scenarios to be prepared.

Following the method outlined above, the analysis will go through the case description in chapter 12 and highlight those parts that are relevant to an analysis of the four abilities.

Central control room moving onshore

A global oil and gas company has decided to move the CCR from offshore to onshore on an existing installation. The company has already in place several expert centres which supports operations, which will continue to support the onshore CCR. The company has for some years integrated contractors closely to operation by use of ISD (Integrated Service Delivery) contracts.

Moving the CCR from offshore to onshore can be assumed to affect the abilities to monitor and to respond, both individually and when coupled. The ability to monitor may be affected because it now has to rely on remote measurements, which clearly only provide a subset of the data available offshore. The ability to respond may also be affected, since fewer resources will be onsite. It is, of course, assumed, that these abilities do not apply to the onshore installation. This assumption may or may not be warranted.

Lean staffing onshore

The company has established lean staffing levels and the installation is remotely operated production part of the time (night). There is a complete offshore control room at the installation, which is manned by one operator on twelve hour day shift. This offshore control room is closed during the night. By only having the offshore CCR manned during the day, the abilities to monitor and to respond will be affected (cf. above). It is reasonable to assume that the most serious consequences will relate to the ability to respond.

Reduced shift schedules, offshore and onshore

The offshore shift schedule is twelve hour dayshift, and fourteen days offshore followed by four weeks at home. The onshore CCR has three shifts, each lasting eight hours. There is an onshore staff rotation every nine months. Operations and planning group staff have standard eight-hour workdays.

A consequence of the shift arrangement and the rotation system is that the teams are not synchronized. This is probably worst in the case of the day shift. A possible consequence of this is that collaboration may suffer, because people do not get to know each other as well as they otherwise could have. This may have consequences for the ability to respond, but is probably more important for the ability to learn, (to share experiences, ways of thinking and operating, and so on).

Distributed support centres; daytime only

The centres support not only this installation, but also the operator's other installations, both at the Norwegian Continental Shelf (seven installations) and partly in other countries. For example, the maintenance support centre, responsible for maintenance planning (opportunity-based maintenance), also supports operations in Africa, the Caspian Sea and the Gulf of Mexico.

The support centres are manned 08.00–16.00 hours local time.

One possible consequence of this has already been mentioned. The fact that the operation is not manned around the clock means that there may be times when it is difficult to get support. This may affect the ability to respond, hence introduce a risk. Furthermore, as the support centres are in different locations and operate in different time zones, it may create uncertainty about who and when to call when the need arises. Another possible risk stemming from distributed (remotely located) support is that knowledge and advice is supplied without a proper understanding of the operational context and the people who requested support.

Integrated contractors

The collaboration is based on the concept integrated contractors. By use of collaboration technology, operators and suppliers are integrated, for example, in morning meetings. The purpose of using an integrated supplier is, among other things, to maximize expertise and to reduce costs associated with the supplier and the operator/oil company having parallel organizations. Central elements in such cooperation are total planning, cross-trained personnel and moving contractor tasks onshore. The contractors share the same real-time information as the operator, and has their own collaboration rooms used for daily communication with the operator, for example, in morning meetings.

This way of organizing and coordinating work among contractors is likely to have unintended effects on both the ability to respond and to anticipate. By 'trimming' the organization, redundancy will inevitably be lost. The capacity of the organization is optimized for normal or usual conditions, but may be insufficient for unusual conditions, and hence affect the ability to respond, particularly if the response is a prolonged one. The focus on 'sharing the same real-time information' means that background information and experience are seen as less important. This may adversely affect the ability to anticipate. Anticipation is not and cannot be a 'regulated' activity, but requires a more 'spontaneous' and flexible combination of ideas, insights and experiences. Anticipation requires diversity within an organization, as well as an organizational climate in which the organization and its management acknowledge the importance of looking ahead.

Cross-disciplinary training

The objective of the cross-disciplinary training across different job categories is to ensure that personnel are trained to carry out tasks in addition to their own field of expertise, even across company boundaries. It has emerged lately that there is a good deal of discontent with many of the new arrivals' level of competence, and that there is concern that events may evolve because those on duty, whether offshore or onshore, will lack the resources to handle the situation.

The purpose of cross-disciplinary training is to trim the organization (make it leaner) without adverse effects on efficiency and readiness. This purpose seems not to have been achieved. The text clearly points to a risk of not being able to respond well enough. In addition, cross-disciplinary training may leave insufficient resources in cases of multiple, simultaneous unexpected events. A further consequence is that the ability to learn may be reduced. Learning is a collective, as much as an individual phenomenon, in the sense that people learn both from observing others and from sharing experiences with others. In order for learning to take place it is also necessary to allow time and provide an opportunity to reflect on what has happened, all of which seems unlikely to be furthered by the principle of cross-disciplinary training.

Recruitment issues

Due to the large numbers of offshore workers retiring in the period 2009–2013 and recruitment problems in the years prior to this period, few people on the platform or in the CCR possess offshore experience of any duration. Domestic recruitment problems are also the reason why a number of offshore engineers have been recruited from countries whose language and culture differ from that of the Norwegian workers. One fears that this development will result in lack of offshore experience at the onshore CCR. One also fears that the onshore operations centre will have problems persuading the staff to rotate to offshore jobs after they have experienced working onshore, resulting in the disintegration of the factory competence. (Note, by the way, that this represents an acknowledged risk rather than a proposed improvement.)

This development points to an important risk, which has been recognized by the company in this example. While the increasing average age of the population is something that no company can do much about, (even if their planning is very long term) the problem of rotating onshore and offshore personnel is clearly relevant. The unwillingness to rotate between positions and the concomitant lack of common experience can easily have adverse effects on the ability to learn and on the ability to respond.

Power supply from onshore

The platform's energy needs are supplied by onshore sources. There is an emergency generator on board for use in the event of power failure. Communication via satellite is also available as a backup in the event of the breakdown of ordinary lines of communication.

The risk of a power failure has clearly been recognized, and reasonable precautions have been taken. Given the general emphasis on cost reduction and leanness, the possibility that the spare capacity is insufficient cannot be ruled out. In other words, the imagined 'worst case' may not be severe enough to match reality. Actual 'worst cases' usually far exceed the bounds of most people's imagination. This may obviously have adverse consequences for the organization's ability to respond.

Discussion

The previous sections have demonstrated how the principles of Resilience Engineering can be used to assess how resilient a specific organization is likely to be. This was done by considering how the abilities to respond, monitor, learn and anticipate would be affected by the organization of work and responsibilities. More precisely, the analysis asked the following questions:

- Will the proposed changes to the organization and to the way it works have any consequences for the organization's ability to respond to expected and/or unexpected events?

- Will the proposed changes to the organization and to the way it works have any consequences for the organization's ability to monitor and detect expected and/or unexpected events?
- Will the proposed changes to the organization and to the way it works have any consequences for the organization's ability to learn from successful and/or unsuccessful responses?
- Will the proposed changes to the organization and to the way it works have any consequences for the organization's ability to think in the long term (anticipate future developments)?

The findings from this analysis are summarized in Table 14.2.

Table 14.2   Summary of resilience-based risk assessment

Proposed IO Changes	Possible Effect on a Resilience Ability			
	Respond	Monitor	Learn	Anticipate
CRC moving onshore	Reduced	Reduced		
Lean staffing offshore	Impaired	Reduced		
Revised shift schedule, offshore and onshore	Reduced		Impaired	
Distributed support centres; daytime only	Impaired			
Integrated contractors	Reduced		(Reduced)	Reduced
Cross-disciplinary training	Reduced		Reduced	
Recruitment issues	Reduced		Reduced	
Power supply from onshore	(Impaired)			
	Brackets indicate the possible effect is uncertain			

This case study describes an organization that clearly places great emphasis on efficiency and 'leanness'. The proposed changes to IO (remote operation) were motivated by studies that showed that 'remote operation of the platform would improve both the efficiency and the safety of the platform; and furthermore extend the life of the field and the wells'. It is, of course, perfectly legitimate for a company to try to improve its efficiency, but it should not be done without taking safety issues into account.

In the present case, the analysis found that many of the proposed changes were likely to adversely affect the organization's resilience and thereby increase its exposure to the consequences of risks and unexpected events.

- The ability to respond was affected by all of the proposed changes. For instance, by moving operations onshore but only manning the offshore CCR during daytime, by distributing support geographically and across time zones, and by relying on integrated contractors.
- The ability to monitor was affected by two of the proposed changes, both having to do with reducing the presence of staff in the offshore CCR.
- The ability to learn was affected by four of the proposed changes, for instance relying on cross-disciplinary training to reduce the number of personnel.
- The ability to anticipate was affected by one of the proposed changes, namely by 'trimming' the supplier and operator organizations.

While the details of the analysis may be debated, and some of the conclusions questioned, it seems clear from the description of the case that the proposed changes have been based solely on their expected benefits in terms of improved efficiency. It has been assumed that safety would not be adversely affected by the proposed changes, or even that it would be improved. But it does not appear that an analysis of safety issues has been carried out, despite the fact that 'the labour organizations opposed the decision of moving the CCR due to safety concerns'. This sanguine attitude is, unfortunately, not unique.

Is the proposed organization suitable for effective safety management?

Keeping in mind the preliminary nature of the present analysis, the answer is unfortunately negative. Taken as a whole, the proposed changes must be judged as detrimental to the readiness and ability of the organization to respond to risks. The proposed organization, and the proposed ways of working, where human resources and competence are distributed geographically, only available part of the time, and where the need to create an effective collaborative environment has been neglected, is far from suitable for effective safety management. It may, of course, be argued that the technical facilities for collaboration have been improved and are state-of-the-art. But technology alone is not sufficient to guarantee effective safety. Furthermore, additional uses of technology usually introduce additional risks. This does not only concern the ability to respond, but also the ability to monitor and to learn.

In terms of risk management, the reduced ability to respond is worrying. However, pointing out this weakness, as well as the weaknesses in monitoring and learning, can form the basis for taking steps to improve the situation. Any of these abilities can be improved, provided they are sufficiently well understood. This can be accomplished by looking at the ability in further detail, for instance, by using the approach described by the Resilience Analysis Grid (RAG) (Hollnagel, 2011b). A more precise description of the functions that are part of the ability and an understanding of how they are coupled to each other can form the basis for developing specific solutions that can be used to compensate for, or overcome the problems.

Does the proposed organization create risks of its own?

Given the limitations of the present analysis, the answer is clearly affirmative. Indeed, the description of the case explicitly mentions several such risks, for instance the difficulty of persuading staff to rotate offshore, and the geographical distribution of vital competencies. Modern technology does indeed make it possible to support 24/7 operations with minimal local staffing, because access can be provided to staff in other locations and time zones. Yet such an arrangement may also make effective support more difficult as important local information and knowledge will be missing. Furthermore, the reliance on advanced IT and communication networks is itself a risk, as we are reminded of on an almost daily basis.

Conclusion

Resilience Engineering is not about assessing accident risks, but about assessing the organization's ability to be resilient in the face of expected and unexpected events. This, of course, includes and requires some idea of the risks, but risk identification is not in itself the main purpose. Resilience Engineering can nevertheless be used to consider or look at specific risks, in this case the risk that the organization will not be able to provide the expected and required functions, and specifically that it will not be resilient (effective) in the face of accidents. The approach described in this chapter has the following characteristics:

- It can be applied to system functions generally, both at the sharp-end (operations) and the blunt end (management). It can be applied during the early stages of system design, when functions may not be known precisely, as well as later, when operations are more mature and better understood.
- It provides an articulated basis for decisions that aim to enhance the system's ability to function in different conditions, that is, the system's resilience. It can be used to guide decisions about resource allocations, that is, how many resources should be given to specific abilities or functions, and when and how they should be applied. In particular it can be used to guide or advise on organizational changes. It can finally be used to decide what the acceptable level of readiness should be.
- The approach assumes that it is possible to estimate the variability of identified system functions, as well as to find the main factors that lead to this variability. The approach also requires that there is adequate expertise available concerning the functions that are being considered.

The approach is a risk assessment of the organization and its safety management system vis-a-vis its resilience abilities, rather than a risk assessment of the direct (or primary) tasks (the work scenario). In other words, it is a risk assessment of the indirect tasks or indirect functions. Direct functions (production) are predicated on the indirect functions (resilience). Knowledge of the risks of the former are necessary to assess the efficacy of the latter, but identifying the risks of the former is not in itself an objective of this analysis.

That said, some fairly obvious risks were identified even at this high level of specification or description. For both production and safety, a functional (or goals-means) decomposition can be performed to determine whether the requisite capabilities are present. The advantage of a Resilience Engineering approach is that it can be used to consider the operational risks even at a high level, such as in the example provided here.

The risk analysis could have been complemented by an analysis of whether any of the suggested changes would lead to improved resilience. Such an analysis would be very much in the spirit of Resilience Engineering and its focus on how things can succeed or go right. To accomplish that would, however, require a different starting point. The questions should focus more precisely on how the organization should be changed to improve its ability to respond, monitor, learn and anticipate. In other words, how to design an organization to improve its resilience? An answer to that question must, however, await another opportunity.

References

AHRQ (Agency for Healthcare Research and Quality). 2007. *Guide to Patient Safety Indicators Version 3.1*. Available at http://tinyurl.com/7e6s7le [accessed: 13 July 2012].

Apneseth, K. 2010. *Resilience in Integrated Planning*. MSc Thesis. Trondheim: NTNU.

Aven, T. 2007. A unified framework for risk and vulnerability analysis covering both safety and security. *Reliability Engineering & System Safety*, 92(6), 745–754.

Aven, T. 2009. Perspectives on risk in a decision-making context – Review and discussion. *Safety Science*, 47(6), 798–806.

Hollnagel, E. 2009. The Four Cornerstones of Resilience Engineering, in *Preparation and Restoration*, edited by C. P. Nemeth, E. Hollnagel and S. Dekker. Aldershot: Ashgate, 117–134.

Hollnagel, E. 2011a. Prologue: The Scope of Resilience Engineering, in *Resilience Engineering in Practice: A Guidebook*, edited by E. Hollnagel, J.Paries, D.D. Woods and J.Wreathall, Farnham: Ashgate.

Hollnagel, E. 2011b. Epilogue: RAG – The Resilience Analysis Grid, in *Resilience Engineering in Practice: A Guidebook*, edited by E. Hollnagel, J.Paries, D.D. Woods and J.Wreathall, Farnham: Ashgate.

ICAO (International Civil Aviation Organization). 2009. *Safety Management Manual*. Doc 9859 AN/474. Montreal, Quebec: ICAO. Available at http://www.icao.int/safety/ism/Guidance%20Materials/DOC_9859_FULL_EN.pdf [accessed 16 July 2012].

Jaubert, J-M. 2006. *The Meaning of Safety*. Available at http://www.offshore-technology.com/features/feature577/ [accessed: 25 June 2010].

Klinke, A. and Renn, O. 2002. A new approach to risk evaluation and management: risk-based, precaution-based, and discourse-based strategies. *Risk Analysis*, 22(6), 1071–1094.

Okrent, D. and Pidgeon, N. 1998. Editorial: Risk perception versus risk analysis. *Reliability Engineering and System Safety*, 59(1), 1–4.

Rosness, R. 2009. A contingency model of decision-making involving risk of accidental loss. *Safety Science*, 47(6), 807–812.

Slovic, P. 2001. The risk game. *Journal of Hazardous Materials*, 86(1-3), 17–24.

Chapter 15
Assessing Risk in Integrated Operations: It's About Choice

Eirik Albrechtsen and Denis Besnard

The previous chapters by Vatn and Hollnagel presented two independent assessments of the same scenario. This chapter compares these two assessments, showing overlaps and differences. We decompose the two approaches in order to help risk managers to identify their strengths and provide them with explicit criteria upon which to base their risk-informed decisions.

Introduction

This chapter follows on from the contributions of Vatn (Chapter 13) and Hollnagel (Chapter 14). In these chapters each author analyses the same Integrated Operations (IO) scenario from their respective standpoint and describes their technical approach to risk assessment. Their approaches are based, respectively, on expressing the uncertainty of future conditions and Resilience Engineering. Both approaches aim to support strategic decision-making and ensure safe operations in the move from an offshore to an onshore control room. Both chapters ended with an assessment of the various types and levels of risk. As it is important to understand the specific characteristics of both approaches, this chapter provides a comparison and a discussion. Each approach can be compared to a house on an island. Rather than forcing them together into a composite method or theory, it makes more sense to build bridges between them. This makes it possible to integrate and link the strengths of each approach.

The chapter will help the reader to understand the differences, compatibilities, strengths and weaknesses of these two approaches to risk assessment. We will therefore attempt to:

- describe and compare the pedigrees of each approach;
- help risk managers to identify potential applications;
- provide high-level recommendations about how to handle risk assessment in IO.

Risk Assessment

Risk assessment is deeply rooted in the offshore oil and gas industry's loss prevention practices. There are different assessment methods for different operational phases and activities. In Chapter 6, Vatn and Haugen distinguish between three types of risk assessment methods: strategic, qualitative design and operative. The risk assessment approaches described in the previous two chapters are examples of strategic analyses that are aimed at developing a safe design and safe operating procedures. Both approaches (expressing uncertainty vs. assessing resilience) provide support for the decision about whether to move an offshore control room onshore.

The first question that arises is whether these two approaches differ from mainstream definitions? An initial comparison can be made with the traditional risk assessment exercise, which provides support for risk-related decision-making. In the traditional risk assessment, risk is typically understood as a combination of the consequences of an event and the likelihood of occurrence. A second comparison can be made with the International Organization for Standardization (ISO) 31000 standard (ISO 31000) that defines risk assessment as the process of:

1. Hazard and threat identification[1] (that is, finding, recognizing and describing sources of risk, events, their causes and their potential consequences).
2. Risk analysis (understanding the nature of risk and determining its level expressed as a combination of consequences and their likelihood).
3. Risk evaluation (comparing the risk analysis with risk criteria to determine whether the risk is acceptable or tolerable).

---

1   ISO 31000 uses the concept of risk identification.

The positions of both Vatn and Hollnagel diverge from this mainstream approach in different ways.

Although Vatn bases his approach on the steps described in the ISO standard, his definition of risk is different. He argues that risk describes uncertainty about the occurrence and severity of events. This contrasts with many other interpretations of risk, which is seen as an inherent property of the system (that is, probabilities or likelihoods are seen as independent of the analyst). On the other hand, Hollnagel does not define risk. Instead, he focuses on how to manage an organization's ability to succeed under varying conditions, which is indirectly related to the level of risk. Unlike the ISO standard, Hollnagel's focus is only on hazard and threat identification. He aims to identify weaknesses in the safety management system that could lead to safety-critical situations.

The Bird's Eye View

Expressing uncertainty (Vatn)

In Chapter 13, Vatn shows how risk can be assessed by following steps that are similar to those of ISO 31000, although with a more of a focus on modelling. However, Vatn departs from the mainstream definition of risk found in the industry. He defines risk as the uncertainty related to the occurrence and the severity of events. This uncertainty can be made operational by expressing the probability of the occurrence of events, together with the severity of potential losses. These evaluations are conditional on communication between stakeholders, the assessor's understanding and assumptions, and verification of the assessment.

As, according to Vatn, risk is about expressing uncertainties the traditional risk assessment also focuses on uncertainties. The first step consists of modelling the primary accident scenario,[2] on the basis of formal logical statements using fault and event trees. In the second step, risk influencing factors (RIFs) are modelled in

2   In a full risk analysis several scenarios would be considered. This was not done in Chapters 13 and 14 as they simply aim to demonstrate the approach.

order to show the impact of human and organizational factors on both events and conditions, and risk itself. RIFs can be modelled at two levels. The first consists of factors that have a direct impact on the event, while second-level factors have an indirect influence (typically on management). This step is followed by change and complexity analyses. It is important to link human and organizational factors to the overall risk picture at this stage, both because it makes it possible to assess the impact of improvement measures on human and organizational conditions, and because the anticipated changes are linked, to a large extent, to changes in human and organizational conditions.

The Resilience Engineering approach (Hollnagel)

Resilience is about succeeding under expected and unexpected conditions (see Chapter 14 for a definition). Therefore, Resilience Engineering is not specifically about risk or about reducing it. Instead, it is about managing the properties of a given system so that it can continue to operate despite disruptions. From this perspective, a low level of risk is a normal side effect of a resilient system and is the reason why Hollnagel pays attention to what he calls 'indirect risks', which are weaknesses in the safety management system. Weaknesses imply insufficient control of hazards and threats, which in turn may lead to accidents.

Four abilities of a resilient system (responding, monitoring, learning and anticipating) are used as a framework to identify the indirect risks in safety management. The first step in Hollnagel's approach is to identify these abilities. The second step is to evaluate how each ability can fail, without analysing the likelihood of occurrence. Therefore, the focus is on the possible consequences of indirect risks, and how they influence system performance.

Comparison of Approaches

These two approaches are now compared according to the following three criteria:

- fundamental assumptions;

- what is assessed;
- how it is assessed.

Fundamental assumptions

Vatn's expression of uncertainty approach (Chapter 13) evaluates whether events will occur in the future and their severity. The current status, together with other information, interpretations and assumptions is used as an input and background to the assessment of future events and scenarios. While Vatn's approach is about judging future conditions, Hollnagel's Resilience Engineering approach (Chapter 14) expresses the current status of the system and how changes will affect it. The 'direct risks' described by Hollnagel are similar to hazards, however the focus of Resilience Engineering is on 'indirect risks', that is, those related to how hazards can be handled.

Therefore, the concept of risk is used very differently in the two approaches. Rather than assessing risk, Resilience Engineering assesses abilities that enable a system to succeed under varying conditions. The assumption is that this capacity can be assessed through an examination of the four structural abilities of an organization (to respond, monitor, learn and anticipate). The level of performance in these four abilities determines the capacity of resilient systems to limit the occurrence and severity of undesired events. Reductions in risk are therefore a side effect of resilience.

Resilience Engineering distinguishes between the classical interpretation of safety (freedom from things going wrong) and the ability to succeed under varying conditions. Nevertheless, it is interesting to note that the Hollnagel's analysis focuses mainly on the negative impacts of the introduction of the onshore control room. He shows how the abilities of a resilient system may be weakened by IO-related changes. Vatn's approach is equally interesting as risk assessment is often said to focus on what goes wrong. However, Vatn suggests that improved coordination capabilities, shared awareness and access to decision-support (for example) brought about by the introduction of IO-related solutions may reduce risk.

Another fundamental difference between the two approaches relates to the sources and limitations of knowledge. Vatn's approach focuses on assessing and expressing uncertainty (that is, lack of knowledge) of future states. Although it is not made explicit in the chapter, his definition of risk indicates that this uncertainty is conditional on dialogue with stakeholders, information, theories, understandings and assumptions, and the verification process. Conversely, Hollnagel does not mention the uncertainty of judgements made by stakeholders.

What is assessed

In general, it could be said that unlike traditional risk assessment, Resilience Engineering is not interested in events, but in organizational properties. From this point of view, risk as such is not evaluated. However, this does not imply that Hollnagel's approach cannot provide support for safety-related decisions and processes. Instead, it means that Resilience Engineering provides an assessment of the structure of an organization. Therefore, it formulates organization-based rather than event-based recommendations. As Hollnagel explains, it is less concerned with failures than system functioning. Moreover, the concept of failure itself is revisited. Instead of treating it as a malfunction (the orthodox approach), Resilience Engineering defines failure as the difficulty of a system to adjust its functioning when faced with varying conditions.

Where Resilience Engineering evaluates organizational properties, uncertainty-based risk assessment essentially expresses uncertainty, including uncertainty related to the ability to succeed. Traditional risk assessment can even be said to offer a broad perspective on all uncertainties regarding undesired events. This raises the question of how risk assessment handles the concept of the ability to succeed under varying conditions. One answer is that all systems are exposed to internal and external demands. How well a system can cope is what matters when expressing uncertainty and finally assessing risk. Whether uncertainty is expressed in terms of success or failure, or whether demands should be called varying conditions are secondary issues.

How it is assessed

Resilience Engineering uses qualitative information in order to produce a semi-quantitative assessment. Although it requires the systematic gathering of specific pieces of data, processing is straightforward. It essentially relies on categorizing data and plotting scores on a set of scales. While data gathering and plotting must be undertaken by specialists, the near-absence of computation has the advantage that the assessment can be understood by the layman. This may, however, be a drawback for risk managers. The focus on properties related to the overall functioning of the system does not produce a numerical risk level that can be associated with, for example, a particular technology. This may be seen as a problem as risk managers often rely on numerical data when taking safety or risk-related decisions.

Uncertainty-based risk assessment involves the structuring and modelling of knowledge (or lack of it) about future events and related uncertainties. Modelling starts with a linear cause and effect structure of fault and event trees. The next step is to model RIFs. Therefore, it is a quantitative approach where modelling is essential. By linking RIFs to fault and event trees, organizational aspects can be introduced into the evaluation. As has already been mentioned, for the expression of uncertainty it is essential to establish the basis for the analysis. This is in contrast to Resilience Engineering, which does not express uncertainty related to subjective judgements.

Bridging Approaches

Following the above comparison of the two approaches the next sections look at how the gaps between them can be bridged and show how they can be applied by a risk manager. Although there is a temptation to merge them into one approach, it cannot be taken for granted that methods or approaches from the same domain can simply be blended. There are multiple differences in terms of technical compatibility, input and output data format, and so on. It is an example of the 'Swiss army knife paradox', whereby the usefulness of a versatile tool diminishes as the specificity of the task increases. Nevertheless, the comparison of the two approaches has shown that there is scope for applying parts of one approach in order to improve the other.

Potential application of Resilience Engineering to uncertainty-based risk assessment

Uncertainty-based risk assessment structures available knowledge to express uncertainty about the occurrence of future events and their severity. In principle, all available knowledge can be applied (there are of course some practical restrictions). Knowledge about the resilience of a system is no exception.

Vatn illustrates how organizational factors can be incorporated into a strategic risk assessment by modelling RIFs. In this way, RIFs for the four Resilience Engineering abilities could be developed: $RIF_{RESPONSE}$; $RIF_{MONITOR}$; RIF_{LEARN}; and $RIF_{ANTICIPATION}$. Similarly, the Resilience Engineering approach can be used to provide input to the assessment of these RIFs, as can the Resilience Analysis Grid (RAG) described in Chapter 8. There is also potential to apply resilience-based thinking into qualitative design and operative risk analyses, for example through the use of checklists similar to those used in HAZard and OPerability Analysis (HAZOP).

Potential application of uncertainty-based risk assessment to Resilience Engineering

Resilience Engineering is about maintaining the four main abilities of a system: to respond, monitor, learn and anticipate. Risk assessment clearly relates to the ability to anticipate future threats and therefore the results of a risk assessment can provide an input to the design of resilient work processes. Operative risk assessments (for example, Job Safety Analyses) are one way to raise the risk awareness of sharp-end operators and can contribute to the anticipation of what can go wrong.

Another potential contribution is the expression of uncertainty used in the traditional risk assessment approach. Vatn suggests that dialogue with stakeholders (which provides information, theories, understandings and assumptions) and verification processes should form the basis for the assessment. This process of consultation and blended data could be used by Resilience Engineering, which as it stands relies essentially on expert, subjective judgements.

Questioning one's choices

An alternative to blending the two approaches is to identify cases and/or questions where one approach offers a better fit than the other. This section lists a series of questions that apply to risk-related decisions, based on Table 15.1.

The purpose of this list is not to encourage a binary choice between uncertainty-based risk assessment and Resilience Engineering. Instead, it draws upon the strengths of each approach to help decision-makers to identify the strong points of scenario-based and structure-based methods, respectively.

- *Objective of the assessment.* Is it important to gather data about the safety-related uncertainty of future operations or about organizational functions?
- *Focus.* Is the focus on operational scenarios or the structure of the organization?
- *Area of application.* Does the assessment concern an operative or design phase?
- *Decision supported.* Is decision support needed for a specific decision (for example, deployment) or is it related to the structure of the organization?
- *Communication.* Will the results of the analysis be looked at by risk specialists or must they be understood by generalists?
- *Compatibility.* Do the new results have to be compatible with, and integrated into existing data?
- *Format.* Is quantitative or semi-quantitative output data required?
- *Expertise needed.* Is there a quantitative risk analysis specialist available? Is there a human factors specialist available?

Table 15.1   Comparison of high-level properties of the two approaches to IO

	Assessment of Risk as Uncertainty Regarding the Occurrence and Severity of Events (Vatn)	Resilience Engineering Approach (Hollnagel)
Main objective	Structure and analyse the uncertainty of the occurrence and severity of events to provide decision support.	Produce knowledge on the performance of system functions, both at the sharp- and blunt-end.
Focus of the analysis	Judgement of future conditions of a system. The analysis focuses on system-specific scenarios. It takes into account organizational dimensions judged relevant by the analyst.	Evaluation of current conditions and how they will be influenced by future changes. The analysis focuses on the properties of the system.
Area of application	Design stage.[1]	Early design stages and operative phases.
Type of decisions supported	Decisions about measures to be implemented.[2] The need to make further investigations and follow-up on issues.	Decisions related to resource allocation, organizational change and accepted level of readiness.
Communication	Requires a well-developed terminology and methods which cannot be described in natural language. Results must be translated into natural language for decision-makers.	Both analysis and communication of results use natural language.
Compatibility	Is close to current industrial tools and practices. Extends existing methods and tools.	Is different from current industrial tools and practices, but not incompatible.
Data format	Quantitative. Operational data must be available. Data on the impact of organizational issues is scarce and expert opinion is required.	Semi-quantitative.
Analyst expertise	Mathematics (modelling) and quantitative risk analysis.	General human factors and organizational resilience.
Relevance to IO	Incorporates risk influencing factors generated by IO-solutions into a risk model.	Assesses how IO-based work processes affect safe and efficient operation.

1   Risk assessments can be used in other stages, but require other methods.
2   Other types of risk assessment methods can support other kinds of decisions; see Chapter 6 for an overview.

The value of interdisciplinary approaches

Although the questions listed above suggest a dichotomy, the point is to trigger further consideration of what should be taken into account before a method-related decision is taken. These questions are far from trivial as the choice made will have an impact on what will be found and what will be done with the results.

Notwithstanding the arguments put forward so far, when it comes to deciding how to analyse risk, there are more than two options. In fact, there are many more decisions to be taken including, what type of risk to analyse, for whom, by whom and from what angle? Another simple question to ask is whether human performance should be included in the risk assessment? This question opens a debate on interdisciplinary approaches and its expected value for managers.

An interdisciplinary approach to a problem can reveal a multitude of facets and lead to optimal solutions. However, it does not happen by accident. Sometimes, the nature of work makes it necessary. To take the example of designing a partly-automated machine for an assembly line; this cannot be done without the involvement of specialists such as automata engineers, health and safety officers, ergonomists, operators and so on. The same should apply to the risk analysis domain. An ideal scenario would include all stakeholders (for example, plant designers and managers, quality, health and safety officers, operators, end-users and so on), and include contributions from all types of specialists (for example, management, safety, human factors, quality, production and marketing).

Conversely, there are obvious costs associated with the approach, notably time and meaning. Producing a risk analysis obviously takes longer if more stakeholders are involved. It can also be difficult for the various disciplines to grasp each other's point of view, leading to the potential for misunderstanding. It might appear that involving personnel from many disciplines in a collaboration boils down to a trade-off: although there are gains they come at a cost. Actually, the argument that the cost of the approach outweighs its benefits only applies in the short term.

In the long run, investment in the production of a deeper and more diverse understanding of risks generates savings in terms of reduced resources consumed by accidents.

Recommendations

The following are some simple risk assessment messages that apply both within and beyond IO, based on the discussions in this chapter.

- Risk assessment of a major organizational change in an IO setting is de facto an exercise that focuses on a complex system. Therefore, it is unlikely that a single method or approach will find answers to all the questions that arise.
- Choosing a risk analysis method itself carries a risk. Asking a few simple questions can lead to the emergence of selection criteria that can form the basis for a discussion of the options.
- Analysing risk from an interdisciplinary perspective has short-term costs but long-term benefits.
- Supplementing standard practice with new methods or learning from other approaches improves safety in the long term.
- The value of risk assessment lies, on the one hand, in how carefully and precisely the system is modelled. On the other hand it can demonstrate uncertainties in the model, analysis and results.

Conclusion

This chapter has compared two methods for assessing risk in the context of IO solutions. We have highlighted how different methods generate different results. We have also discussed how the choice of a method to analyse safety-related issues can be based on the answers to a few simple questions. It is important to explore these issues as they have a direct effect on fundamental issues such as why and for whom the risk analysis is prepared. Finally, we hope that these examples can provide a starting point for reflection on the added value that diversity brings to the evaluation of risk.

References

ISO 31000, *ISO 31000:2009. Risk Management – Principles and Guidelines*.

Chapter 16
Lessons Learned and Recommendations from Section Three

Denis Besnard and Eirik Albrechtsen

Take-home Messages from Chapters 12–15

In Chapter 12 Albrechtsen documented an Integrated Operations (IO) scenario that provided the basis for a comparison of two approaches to risk assessment.

In Chapter 13 Vatn approached the problem described in the scenario from a quantitative risk assessment perspective. The main operational messages are as follows:

- Risk can be assessed by expressing uncertainty regarding the occurrence and severity of events.
- Risk must also include such dimensions as the level of dialogue between stakeholders, the assessor's understanding of the system being assessed and the performance of the verification process.
- Other, related aspects must be taken into account such as risk influencing factors (RIFs), the effects of change and complexity.

Hollnagel (Chapter 14) also tackled the problem described in the scenario, but from a Resilience Engineering perspective. The main operational messages are as follows:

- In terms of safety management, Resilience Engineering is

about the performance of system functions that allow it to be safe (as opposed to avoiding being unsafe).

- Resilience Engineering rests on four essential abilities: knowing what to do, knowing what to look for, knowing what has happened and knowing what to expect. Each of the four abilities can be evaluated and given a score. The aggregated scores provide a consolidated resilience assessment of the system in question.
- Reducing and coping with risk is a normal side effect of a resilient system. Assessing resilience can therefore be linked to risk assessment.

Albrechtsen and Besnard (Chapter 15) compared the two risk assessment methods described in Chapters 13 and 14. The main operational messages are as follows:

- There are both differences and links between the various methods for assessing risk and the safety performance of a system.
- An alternative to blending approaches is to identify cases and/or questions where one approach offers a better fit than the other.
- A diversity of assessment methods adds value to the understanding of risks.

Recommendations

On the basis of the chapters summarized above, the following operational recommendations can be made:

- There are many methods for assessing risks, safety and resilience. The selection of the most appropriate method should be based on the nature of the system in question.
- One useful selection criterion could be the role played by humans in the system. Will they be part of it? Will they interact with it? Will they be independent of it? Another potentially useful criterion is granularity. Will the system operate in a sand-box or could it potentially affect the organization?
- These points give rise to the question of how a company

selects the method(s) it will use. For example, is the choice culture-driven ("This is the method we use here.") or is it needs-driven ("How shall we handle this assessment given the system in question.")?

- Choosing more than one assessment method makes sense for complex socio-technical systems. One option is to use similar methods in order to consolidate results through redundancy. Although this might narrow the focus of the analysis, it might also increase confidence in the output. Another option is to use different methods in order to consolidate results through diversity. This will increase the breadth of the output at the cost of making it more challenging to integrate the results.

Chapter 17
Managing Risks in Integrated Operations: Interdisciplinary Assessments beyond Technology and Methods

Denis Besnard, Eirik Albrechtsen and Jan Hovden

In this conclusion we will address some overarching issues that were difficult to address in individual chapters but nonetheless have an impact on the practice of risk assessment, namely interdisciplinarity, trade-offs and change management.

Introducing new work processes in an Integrated Operations (IO) environment triggers important changes and managing them can be challenging. These changes can have both a positive and negative impact on major accident prevention. On the positive side, IO solutions can, for example, provide improved access to risk-related information, improved communication and collaboration across disciplines and organizations, and better ways of presenting safety-critical information. IO can also pose a challenge for major accident prevention because of, for example, the complexity of organizing work and an inadequate flow of information.

Another challenge is to identify hazards and vulnerabilities, assess risk and evaluate the safety performance within these new ways of organizing work. More precisely, a critical issue is the choice of what to assess and how, given the magnitude of the change that IO brings about. This is the angle taken throughout this book, which offered some answers.

The Variety of Assessment Approaches

Table 17.1 is a guide to the tools, methods and reasoning frameworks presented in the book. It aims to help practitioners and managers to decide when, how and why they should use the different approaches.

Table 17.1   A guideline for applying the methods, tools and reasoning frameworks presented in the book

Book Chapter and Author(s)	What Was Presented in the Chapter	Why Use It?	How to Use It?	Why Is It Relevant to IO?
2. IO concepts and their impacts on major accident prevention (Besnard and Albrechtsen)	A *reasoning framework* that shows a) how IO solutions improve assessment and management of risk and b) new issues that must be considered in risk assessment in an IO context.	To understand the core aspects of IO and to understand the link between different elements in risk management and how these are linked to IO concepts.	As a conceptual model to understand IO and how IO influences risk assessment and management.	Indicates how the various IO concepts and their integration influence major accident prevention both in a positive and negative direction.
3. Using human and organizational factors to handle the risk of a major accident in IO (Andersen)	A *reasoning framework* that identifies *operative* IO-based human and organizational factors that may increase major accident risks.	To understand how human and organizational factors may increase major accident risks.	As a conceptual description of what may go wrong in an IO environment.	Shows how specific IO concepts create safety challenges
	A *tool* (a set of diagnostic questions) to assess human and organizational factors in an *operative* IO context.	To assess safety issues in terms of what can go wrong related to human and organizational factors in IO-based drilling operations.	As a list of IO-relevant questions to diagnose safety. It can be used in various safety assessment methods (e.g. risk analysis, audits, accident investigation).	Provides a list of questions to study IO-based impacts on risk and safety. The list can be modified and applied to other activities than drilling.
4. Assessing risks in systems operating in complex and dynamic environments (Grøtan)	A *reasoning framework* that identifies the signs of different types of complexity, relevant to *design, planning* and *operation*.	To understand and identify various complex and dynamic issues.	As a conceptual framework for a priori identification and understanding of complexity.	Increases awareness of how IO can produce complex situations, e.g. many actors involved with different agendas.

Book Chapter and Author(s)	What Was Presented in the Chapter	Why Use It?	How to Use It?	Why Is It Relevant to IO?
6. On the usefulness of risk analysis in the light of Deepwater Horizon and Gullfaks C (Vatn and Haugen)	A *reasoning framework* that identifies deficiencies in current risk analysis practices related to both *design* and *operation*.	To understand some of the weaknesses of risk analysis and how it can be improved.	As an input to improving the quality of risk analysis.	Shows how IO solutions can improve operative risk analyses.
7. Assessing the performance of human–machine interaction in eDrilling operations (Besnard)	A *reasoning framework* identifying potential human–machine pitfalls in *design* and *operation*.	To understand potential pitfalls in human–machine interaction for highly automated tasks.	As a conceptual description of what may go wrong in human–machine interaction in an IO environment.	IO solutions generate automated tasks as well as remote control and supervision, which create the possibility of human–machine interaction failures.
	A *method* (including a set of diagnostic questions) to assess the *operative* performance human–machine interaction in drilling-related control and supervision tasks.	To assess potential pitfalls in human–machine interaction for highly automated tasks.	As a tool to assign scores to elements of human–machine interaction, which are aggregated into an overall score.	
8. Measuring resilience in integrated planning (Apneseth et al.)	A *method* (including a checklist) for assessing the resilience of integrated planning.	To assess the resilience of integrated planning processes.	As a tool to evaluate critical functions that must deliver wanted outcomes. Scores are assigned to these functions to express the level of resilience.	Ensuring adequate resilient performance is necessary for complex socio-technical systems such as IO. The method can be adjusted to other IO-based activities.
9. Resilient planning of modification projects in high risk systems: The implications of using the FRAM for risk assessments (Tveiten)	A *method* for assessing latent failures and potential variability that can lead to unwanted outcomes in the *planning* phase of modification projects.	To analyse unwanted consequences due to performance variability in order to develop countermeasures to dampen performance variability.	As a framework to model and describe functions for evaluating potential variability and identify functional resonance.	The method is well-suited for studying functions in complex socio-technical systems such as IO.

Table 17.1   *concluded*

Book Chapter and Author(s)	What Was Presented in the Chapter	Why Use It?	How to Use It?	Why Is It Relevant to IO?
10. Promoting safer decisions in future collaboration environments – mapping of information and knowledge onto a shared surface to improve onshore planner's hazard identification. (Rindahl et al.)	A *tool* for visualizing risk-related information in the *planning* of maintenance operations.	To present risk-related information by use of visualization technology.	As a prototype for visualizing risk-related information in a collaborative environment.	Demonstrates how IO solutions (collaboration and visualization technology) can improve risk management.
13. Steps and principles for assessing and expressing risk in an IO setting (Vatn)	A *method* for assessing the impact of change on major accident risk by expressing the uncertainty of occurrence and severity of events.	To assess uncertainty regarding the occurrence and severity of unwanted events due to changes.	As a set of steps to identify, analyse and evaluate risk and express it as the uncertainty of occurrence and severity of events.	To manage the change process that IO implies.
14. A Resilience Engineering approach to assess major accident risks (Hollnagel)	A *method* for assessing the impact of change on a system's resilience and how this influences risk.	To assess how change impacts the performance of safety management decisions, which in turn influence risk.	As a survey-driven analysis of the impact of change.	Ensuring an adequate resilient performance is necessary for complex socio-technical systems such as IO.
15. Assessing risk in IO: It's about choice (Albrechtsen and Besnard)	A *reasoning framework* to bridge resilient-based and uncertainty-based assessments of risk.	To understand why and how multiple assessment approaches should be integrated.	As an example of how different approaches to, and perspectives on risk can be aligned.	IO is a socio-technical change that requires an interdisciplinary approach to understand, assess and manage safety and risk.

Executing Projects with Several Contributing Disciplines

The first chapter of the book states that an interdisciplinary approach is suited both to understanding hazards and threats and assessing risk related to IO. This book demonstrates the main results of the interdisciplinary approach taken by the RIO research project.[1] The RIO project aimed to address risk related to the new ways of organizing work brought about by the IO context and specifically, how to assess and express risk in such a context.

Because there were several disciplines involved, the project itself had to address the issues of collaboration – and it was far from trivial to achieve. We believe the issue is relevant to many projects other than academic research. Many, if not all sectors of the industry are in constant need of collaboration between members of staff from various in-house disciplines. The problem exists throughout the entire lifecycle of projects and involves many different stakeholders, from procurement and design to delivery and deployment. As the issue of collaboration cuts across so many fields, we would like to share our own thoughts and experience. That is the subject of this section.

Inter- or multidisciplinarity

In hindsight, the interdisciplinary aim of the project clearly developed into a multidisciplinary approach. Although several attempts were made to a) cross traditional boundaries between disciplines and schools of thought and b) integrate knowledge from different disciplines, it proved to be a difficult task. Overall, the project and this book (with the exception of Chapter 15) ended up being the result of contributions from a combination of disciplines, that is, a multidisciplinary collaboration. The value of the approach lies is the range of perspectives it offers, however, our intention to integrate these perspectives will have to be left to others.

Various attempts were made to arrive at an interdisciplinary approach. Our efforts included a series of workshops, seminars with the industry, a collection of essays and memos, a 'live' discussion document that every project member could edit, the scenario exercise described in Chapter 12 and email discussions. Despite these attempts, the project struggled to integrate knowledge through interactions between disciplines. While Albrechtsen et al. (2010) go into more detail on these issues, the essence of their work is presented here.

---

1   The full name of the research project is 'Interdisciplinary Risk Assessment in Integrated Operations addressing Human and Organizational Factors'.

It could be said that the title of the project ('Risk Assessment in Integrated Operations') gave priority to traditional risk analysis, which is seen to provide direct risk information to decision-makers. However, some contributors hardly addressed the concept of risk at all. Instead they sought to identify relevant properties of socio-technical systems in order to evaluate safety performance, risks and/or vulnerabilities. Academics from various backgrounds including, for example, Resilience Engineering and complexity theory were able to comment on the remit and shortcomings of risk analyses. Nevertheless, their comments were perceived to be unfair by risk analysts, who thought that the concept of risk and its assessment had not been properly understood.

Another major challenge for participants was to try to understand each other's point of view and terminology especially in relation to risk, uncertainty, complexity and intractability. For instance, there were at least three different interpretations of risk in the project, namely:

- as a system property involving probability and consequences;
- as an expression of uncertainty about an occurrence and its consequences as expressed by an analyst; and
- as system properties related to threats and hazards (threats and hazards were often used interchangeably).

To a certain extent, the project teamwork was a mixture of 'trench warfare', striving to be understood and accepted by colleagues from other disciplines and maintaining an open mind when listening to the contributions of others.

Lessons learned for future interdisciplinary attempts

Based on our experiences of working on the RIO project (together with the work of Albrechtsen et al., 2010), we would like to offer some advice for any future interdisciplinary safety-related research projects, whether in industry or academia:

- At the beginning of the project explicitly establish different definitions and understanding of key concepts. Respect these different interpretations. Participants must be able to use different definitions as long as it is made clear which one is being used.

- A shared, tangible problem will help project members to work towards the same objectives.
- Remember that others are experts too. Different skills mean that the world will be seen from various perspectives and framed differently.
- Take the fact that project members are from different disciplines seriously. Look for differences and establish a platform that enables contact and communication. Highlight the differences and communicate about diversity both internally (within project teams) and externally (to stakeholders).
- Cultivate what you have in common as well as the differences.
- Constructive criticism of the approach taken by another discipline can be useful, but should be given carefully. Show a) the rationale behind the comments, and b) how all parties can benefit from a debate.
- Push participants to leave their comfort zone, and take a journey into unfamiliar academic waters. Exposure to new ideas can expand your perspective and encourage you to think twice before dismissing unfamiliar concepts.
- Be constructive: help each other, share stories. Ask yourself how your knowledge can help others.
- Disciplines consist of individuals. Therefore, interdisciplinarity is not about collaboration between disciplines, it is about collaboration between people. How individuals get along on a personal level matters.
- Power games (over models, agendas and control) should be expected and have to be dealt with. Project managers must be able to control the situation, as well as the resulting personal conflicts.

Some Elements of a Risk Assessment Approach

The nature of data and related beliefs

Risk assessment methods have been strongly influenced by engineering practices. They are rooted in all phases of the lifecycle of an oil and gas installation, ranging from quantitative design assessments to qualitative, operative assessments. This means that any contributions from other domains are seen as support functions that feed the risk assessment. This situation suggests that quantitative risk assessment is the solution to all risk problems. However, experts from other domains see the quantitative assessment of risk or human performance as an oversimplification of reality. These beliefs are strong barriers to interdisciplinarity. In fact, it is a type of scientific fundamentalism – a distorted belief in the usefulness of one's own discipline, whereas scientists working in safety and risk should be following an interdisciplinary approach.

In the industrial world, skills and resources (for example, time and finances) are limited. A comprehensive, fully interdisciplinary, flawless risk assessment is not a viable option for them. Choices have to be made. Given the real operational constraints, academic views on risk assessment practices may seem naïve. However, our point is not to prescribe new practices. Instead, we argue that in risk assessment – as in many other safety-related decisions – the trade-off between resources and objectives has to take into account the variety of approaches and disciplines available. This variety includes the nature of the output data and changing one's beliefs. For instance, it is not necessarily an impediment that part of a risk assessment cannot be delivered in a quantitative form. In fact, the real impediment would be to discard qualitative data altogether. The assumption that risk-related decisions must be taken on the basis of numbers can make managers oblivious to crucial risk-related issues. The obstacle is clearly beliefs.

Many risks but one method?

A particular bias that we would like to address is that of only using one method or approach to assess any type of risk or take any kind of decision. Even if it is appealing in the short term (for example, because it saves resources), the long-term effects can be counterproductive for safety. Any method or approach will only capture what it was designed for and most likely miss everything else. Moreover, it should not be forgotten that just because a risk assessment exercise has been conducted does not mean that subsequent decisions will be better informed than before. This is an issue that relates to the diversity of competences held by risk assessors. For example, during a technological change, the engineer will typically focus the risk assessment on technical failures or physical hazards. On the other hand, human factors experts will attempt to understand dimensions such as the human impact of change. From a scientific point of view, what makes most sense is a combination of the two approaches rather than a choice of one or the other.

The evolution of new methods and integration with those already in use is another issue. The new challenges that have emerged with the introduction of IO do not mean that what used to be standard risk assessment practices must be abandoned. Instead, a greater diversity of techniques can complement those already in use. For example, Resilience Engineering (see Chapter 14) focuses on four organizational capacities in order to assess the safety performance of system functions, which may indirectly affect risk. This type of approach operates at the level of the company or plant. It complements technical assessments such as fault tree analyses, which tend to focus on physical equipment. The lesson to be learned from this is that what might have been thought of as an exclusive choice between two options is actually one of integration.

Many methods, limited resources

Readers will not be surprised to read that there are many risk assessment methods (see, for example, Mohaghegh et al. 2009 for an overview). However, the choice of method is critical. In an ideal world, the decision should vary according to the scenario in question, the level of granularity at which the analysis must be conducted, and so on. In fact, the main selection criteria are resources such as competences and time. (Besnard and Hollnagel, 2012). The choice of method is guided by operational constraints rather than the needs of the assessment in question. Some other biases include:

- selecting the method that the company uses by default;
- lack of competences to identify the need for a different method;
- no time or budget to allocate to learning a new assessment technique;

- the regulatory body does not require the use of any particular method.

The consequence of these biases on risk management includes overlooking certain types of risks (for example, organizational) because the skills to identify them are lacking. Moreover, the risks that are analysed may be limited to those that the chosen method is designed to process (see Lundberg et al., 2009). Risk managers end up in a situation where the resources (finances, time and so on) required to master a variety of methods are implicitly traded-off against the perceived return on investment (safety performance).

Whatever the task, there are never enough resources to do things perfectly. Therefore, trade-offs are unavoidable and they inevitably impact safety (Hollnagel, 2009). However, simply blaming trade-offs for failure would miss the point entirely. Instead, it is important to acknowledge the existence of trade-offs in safety-critical areas and to examine the organization and its operational context to find their origins.

What place left for humans in risk assessment?

One issue has been hardly mentioned in this book: how should new risk assessment techniques be managed? This implicitly raises the question of how to ensure that risk assessment evolves alongside the system to be assessed. How should new risk assessment practices that are imposed by technological change be managed? The enforcement of a new, single method to replace the old one cannot work. It will simply lead to workarounds by members of staff who try to continue to do what they know. Instead, change must be facilitated. To this end, some basic conditions must be present:

- Senior management must define and agree on where they want to go with risk assessment as a means to manage safety (for example, targeted safety performance, the extent to which humans should be included in assessments, the range of dimensions measured, the spectrum of candidate methods).

- This strategic vision must be backed up by corresponding resources (for example, training, hiring new staff, purchasing licenses, subcontracting).
- Experts and practitioners must be consulted (for example, What new methods are being considered? What is the practitioners' interest in this change? Is this change feasible, how long will it take and what will the human cost be?).
- A progressive and collaborative deployment plan must be developed (for example, Who will pilot change? Which changes must happen first? What are the test cases? When does testing end?).
- A set of indicators must be determined to measure the effects of change (for example, including the human cost of change, gains in safety performance).
- A long-term monitoring scheme must be in place from the outset to evaluate the match between the strategic vision, the results and the cost of producing them.

The scope of the book did not allow for a full chapter on change management. We have tried to compensate for this limitation here. The above list of bullet points is a starting point for thinking about what areas of change must be managed and how.

On a more general note, a change of technology always has a human counterpart, which can be captured by a human factors or organizational science specialist. Unfortunately, how change actually reshapes the working practices of an individual is generally overlooked or misunderstood. There are several reasons for this. One is a techno-centric management style, which is clearly detrimental to systems. It assumes that technology can be adopted or deployed independently of its end-users or the surrounding organization. After all, it is cheaper to do so. Or is it? Actually the opposite might be true. In the long run, deploying a technology or method without assessing the potential human consequences may turn out to be vastly unproductive. At best, it will result in workarounds and make strategic objectives very difficult to achieve. This applies to the deployment of IO; it also applies to choosing the next generation or combination of methods to assess its related risks. These decisions, although technical, can only be profitable if the most important system component is given the central role. This component is definitely not technology.

References

Albrechtsen, E., Hovden, J., Størseth, F. and Vatn, J. 2010. *Interdisciplinary Risk and Safety Research: Trench Warfare, Power Plays and Humpty Dumpty*. Paper presented at Workingonsafety.net, Røros, Norway, 7–10 September 2010.

Besnard, D. and Hollnagel, E. 2012. I want to believe. Some myths about the management of industrial safety. *Cognition, Technology & Work* (online access only as of June 23rd, 2013; DOI 10.1007/s10111-012-0237-4).

Hollnagel, E. 2009. *The ETTO principle: Efficiency-Thoroughness Trade-Off: Why Things That Go Right Sometimes Go Wrong*. Aldershot: Ashgate.

Lundberg, J., Rollenhagen, C. and Hollnagel, E. 2009. What-You-Look-For-Is-What-You-Find. The consequences of underlying accident models in eight accident investigation manuals. *Safety Science*, 47(10), 1297–1311.

Mohaghegh, Z., Kazemi, R. and Mosleh, A. 2009. Incorporating organizational factors into Probabilistic Risk Assessment (PORA) of complex socio-technical systems: A hybrid technique formalization. *Reliability Engineering and System Safety*, 94(5), 1000–1018.

Index